Simplified Design of Structural Wood

#17509018

Simplified Design of Structural Wood

The Late Harry Parker, M.S.

Formerly Professor of Architectural Construction
University of Pennsylvania

FOURTH EDITION

prepared by

James Ambrose, M.S.

Professor of Architecture
University of Southern California

WILEY

A Wiley-Interscience Publication

JOHN WILEY & SONS

New York Chichester Brisbane Toronto Singapore

Library of Congress Cataloging-in-Publication Data

Parker, Harry, 1887–
Simplified design of structural wood.

"A Wiley-Interscience publication."
1. Building, Wooden. 2. Structural design.
I. Ambrose, James E. II. Title.
TH1101.P37 1988 694'.1 88-154
ISBN 0-471-85134-5

Printed in the United States of America

10 9 8 7 6 5 4 3 2 1

Preface to the Fourth Edition

The publication of this edition of Professor Parker's enduringly popular book asserts the position that the need for such a treatment of the subject continues to exist. In the face of ever-increasing sophistication and complexity of work in structural engineering, the previous editions of this book have provided a useful opportunity for learning for those persons with limited training in mathematics and engineering analysis. The purpose of this book is to discuss the common, frequently encountered problems relating to design of components and systems of wood for building structures. In general, computational work is limited to the use of high school arithmetic, algebra, and geometry. It is also generally assumed that the reader has no previous experience in engineering investigation or design.

This edition contains essentially all of the materials treated in the previous edition, although some topics or methods of construction now less frequently encountered have been eliminated or reduced in scope. Data and criteria have been brought into conformity to the latest references following Chapter 20. Since most of the references still use primarily U.S. units of measurement, the work in this edition is primarily presented in that system, although equivalent results and data are frequently given in SI metric units.

Considerable new material has been added in this edition, including treatments of pole structures, joints using nails and screws, mechanically driven fasteners, plywood gussets, manufactured trusses, and wood fiber products. A major new chapter—Chapter 19—deals with plywood-sheathed horizontal diaphragms and vertical shear walls; the lateral bracing system is used for the majority of wood building structures. A second new chapter—Chapter 20—presents illustrations of the complete design of wood structural systems for two example buildings.

In preparing this edition I have endeavored to retain the spirit and basic aims of the previous editions, as stated in Professor Parker's preface to the first edition which follows. Although industry and design practices have changed considerably, the issues, needs, and the audience addressed by Professor Parker remain essentially the same.

I am grateful to the National Forest Products Association, the International Conference of Building Officials, and John Wiley & Sons for permission to use materials from their publications. I am also highly appreciative of the efforts of the editors and production personnel at John Wiley in New York for their diligence in turning my rough cuts into a fine presentation. This work would not have been possible without the support and encouragement of my colleagues in the School of Architecture at the University of Southern California and the tolerance, patience, and help of my family. I must also acknowledge a considerable respect for the contributions made by the preparer of the third edition of this book, my good friend, Harold Dana Hauf.

James Ambrose

Westlake Village, California
May 1988

Preface to the First Edition

This volume is the fifth of a series of elementary books relating to the design of structural members in buildings. The author has endeavored to explain as simply as possible the methods commonly used in determining proper timber sizes. This book deals primarily with wood members that support loads in buildings.

The material is arranged not only for classroom use but also for the many young architects and builders who desire a guide for home study. With this in mind, a major portion of the book is devoted to the solution of practical problems, which are followed by problems to be solved by the student. In addition to explanations of basic principles of mechanics involved in the design of members, numerous safe load tables have been included. These tables will enable one to select promptly members of proper size to use for given conditions.

Stress tables, properties of sections, and the tabulations of technical information pertinent to timber construction are included, thus making reference books unnecessary.

It is assumed that those who use this book have had no previous training. As in the previous books of the series, the use of advanced mathematics has been avoided, and a knowledge of high school arithmetic and algebra is all that is necessary to understand the mathematics involved.

In preparing material the author has employed the commonly used design procedures. He has drawn freely from recommendations and suggestions advanced by leading authorities on timber construction, namely, the Forest Products Laboratory of the United States Department of Agriculture, the National Lumber Manufacturers Association, the Timber Engineering Company, the Southern Pine Association, the West Coast Lumbermen's Association, and the American Institute of Steel Construction. Grateful acknowledgment is made to these associations and agencies for their kindness in granting permission to reproduce tables and technical information. Without this cooperation a book of this nature would be impossible.

HARRY PARKER

High Hollow, Southampton
Bucks County, Pennsylvania
1948

Contents

Introduction

The following suggestions are offered as a guide to the effective use of this book in acquiring a working knowledge of the design of wood structures.

1. Take up each item in the sequence presented and be certain that each is thoroughly understood before continuing with the next.
2. Since each problem to be solved is prepared to illustrate some basic principle or procedure, read it carefully and make sure you understand exactly what is wanted before starting to solve it.
3. Whenever possible make a sketch showing the conditions and record the data given. Such diagrams frequently show at a glance the problem to be solved and the procedure necessary for its solution.
4. Make a habit of checking your answers to problems. Confidence in the accuracy of one's computations is best gained by self-checking. However, in order to provide an occasional outside check, answers to some of the exercise problems are given at the end of the book. Such problems are indicated by an asterisk (*) following the problem number where it occurs in the text.
5. In solving problems, form the habit of writing the denomination of each quantity. The solution of an equation will be a number. It may be so many pounds, or is it pounds per

square inch? Are the units foot-pounds or inch-pounds? Adding the names of the quantities signifies an exact knowledge of the quantity and frequently prevents subsequent errors. Abbreviations are commonly used for this purpose, and those employed in this book are identified in the following discussion of units of measurement.

Computations

In professional design firms structural computations are most commonly done with computers, particularly when the work is complex or repetitive. Anyone aspiring to participation in professional design work is advised to acquire the background and experience necessary to the application of computer-aided techniques. The computational work in this book is simple and can be performed easily with a pocket calculator. The reader who has not already done so is advised to obtain one. The "scientific" type with eight-digit capacity is quite sufficient.

For the most part, structural computations can be rounded off. Accuracy beyond the third place is seldom significant, and this is the level used in this work. In some examples more accuracy is carried in early stages of the computation to ensure the desired degree in the final answer. All the work in this book, however, was performed on an eight-digit pocket calculator.

Symbols

The following "shorthand" symbols are frequently used.

Symbol	Reading
$>$	is greater than
$<$	is less than
$\geq$	equal to or greater than
$\leq$	equal to or less than
6′	6 feet
6″	6 inches
Σ	the sum of
ΔL	change in L

TABLE 1. Units of Measurement: U.S. System

Name of Unit	Abbreviation	Use
	Length	
Foot	ft	Large dimensions, building plans, beam spans
Inch	in.	Small dimensions, size of member cross sections
	Area	
Square feet	ft^2	Large areas
Square inches	$in.^2$	Small areas, properties of cross sections
	Volume	
Cubic feet	ft^3	Large volumes, quantities of materials
Cubic inches	$in.^3$	Small volumes
	Force, Mass	
Pound	lb	Specific weight, force, load
Kip	k	1000 lb
Pounds per foot	lb/ft	Linear load (as on a beam)
Kips per foot	k/ft	Linear load (as on a beam)
Pounds per square foot	lb/ft^2, psf	Distributed load on a surface
Kips per square foot	k/ft^2, ksf	Distributed load on a surface
Pounds per cubic foot	lb/ft^3, pcf	Relative density, weight
	Moment	
Foot-pounds	ft-lb	Rotational or bending moment
Inch-pounds	in.-lb	Rotational or bending moment
Kip-feet	kip-ft	Rotational or bending moment
Kip-inches	kip-in.	Rotational or bending moment
	Stress	
Pounds per square foot	lb/ft^2, psf	Soil pressure
Pounds per square inch	$lb/in.^2$, psi	Stresses in structures
Kips per square foot	$kips\text{-}ft^2$, ksf	Soil pressure
Kips per square inch	$kips\text{-}in.^2$, ksi	Stresses in structures
	Temperature	
Degree Fahrenheit	°F	Temperature

TABLE 2. Units of Measurement: SI System

Name of Unit	Abbreviation	Use
	Length	
Meter	m	Large dimensions, building plans, beam spans
Millimeter	mm	Small dimensions, size of member cross sections
	Area	
Square meters	m^2	Large areas
Square millimeters	mm^2	Small ares, properties of cross sections
	Volume	
Cubic meters	m^3	Large volumes
Cubic millimeters	mm^3	Small volumes
	Mass	
Kilogram	kg	Mass of materials (equivalent to weight in U.S. system)
Kilograms per cubic meter	kg/m^3	Density
	Force (Load on Structures)	
Newton	N	Force or load
Kilonewton	kN	1000 newtons
	Stress	
Pascal	Pa	Stress or pressure (one pascal = one N/m^2)
Kilopascal	kPa	1000 pascals
Megapascal	MPa	1,000,000 pascals
Gigapascal	GPa	1,000,000,000 pascals
	Temperature	
Degree Celsius	°C	Temperature

TABLE 3. Factors for Conversion of Units

To Convert from U.S. Units to SI Units Multiply by	U.S. Unit	SI Unit	To Convert from SI Units to U.S. Units Multiply by
25.4	in.	mm	0.03937
0.3048	ft	m	3.281
645.2	in.^{2}	mm^2	1.550×10^{-3}
16.39×10^3	in.^{3}	mm^3	61.02×10^{-6}
416.2×10^3	in.^{4}	mm^4	2.403×10^{-6}
0.09290	ft^2	m^2	10.76
0.02832	ft^3	m^3	35.31
0.4536	lb (mass)	kg	2.205
4.448	lb (force)	N	0.2248
4.448	kip (force)	kN	0.2248
1.356	ft-lb (moment)	N-m	0.7376
1.356	kip-ft (moment)	kN-m	0.7376
1.488	lb/ft (mass)	kg/m	0.6720
14.59	lb/ft (load)	N/m	0.06853
14.59	kips/ft (load)	kN/m	0.06853
6.895	psi (stress)	kPa	0.1450
6.895	ksi (stress)	MPa	0.1450
0.04788	psf (load or pressure)	kPa	20.93
47.88	ksf (load or pressure)	kPa	0.02093
0.566 × (°F − 32)	°F	°C	(1.8 × °C) + 32

Units of Measurement

At the time of preparation of this edition, the building industry in the United States is still in a state of confused transition from the use of English units (feet, pounds, etc.) to the new metric-based system referred to as the SI units (for Système International). Although a complete phase-over to SI units seems inevitable, at the time of this writing the construction materials and products suppliers in the United States are still resisting it. Consequently, most building codes and other widely used references are still in the old units. (The old system is now more appropriately called the U.S. system because England no longer uses it!) Although it results in some degree of clumsiness in the work, we have chosen to give the data and computations in this book in both units as much as is practicable. The technique is generally to perform the work in U.S. units and immediately follow it with the equivalent work in SI units enclosed in brackets [thus] for separation and identity.

Table 1 lists the standard units of measurement in the U.S. system with the abbreviations used in this work and a description of the type of the use in structural work. In similar form Table 2 gives the corresponding units in the SI system. The conversion units used in shifting from one system to the other are given in Table 3.

Standard Notation

The following notation is used in this book and is in general conformance to that used in most of the references.

a	1. Moment arm; 2. Increment of an area
A	Gross (total) area of a surface or a cross section
b	Width of a beam cross section
c	Distance from the neutral axis to the edge of a beam cross section
C_f	Form factor
C_F	Size factor
C_s	Slenderness factor for bending member
d	Depth of a beam cross section or overall depth of a truss
D	1. Diameter; 2. Deflection
e	1. Eccentricity (dimension of the mislocation of a load resultant from the neutral axis, centroid, or simple center of the loaded object); 2. Unit elongation
E	Modulus of elasticity for tension, compression, flexure
f	1. Computed stress; 2. Frequency
F	1. Force; 2. Limiting or allowable stress
F_b	Design value for bending stress
F_c	Design value for compression stress parallel to the grain
$F_{c\perp}$	Design value for compression stress perpendicular to the grain
F_n	Design value for compression stress at an angle to the grain

F_t	Design value for tension stress parallel to the grain
F_v	Design value for horizontal shear stress
F'_c	Design value for compression stress parallel to the grain, adjusted for column slenderness effect
G	Specific gravity
h	Height
H	Horizontal component of a force
I	Moment of inertia
J	1. Polar moment of inertia; 2. Factor used in column interaction
K	Empirical coefficient for allowable compression stress in solid columns of intermediate slenderness
K_e	Effective buckling length factor
l	Length (usually in inches)
L	Length (usually in feet)
M	1. Moment of a force; 2. Magnitude of internal bending moment in a beam
n	Ratio of the moduli of elasticity of two interacting materials
N	Number of
p	1. Percent; 2. Unit pressure
P	1. Concentrated load (force at a point); 2. Allowable load for a fastener in direction parallel to the grain
q	Unit of a uniformly distributed linear load
Q	Allowable load for a fastener in direction perpendicular to the grain
r	Radius of gyration
R	Radius of curvature (of a circle, etc.)
s	1. Center–to–center spacing of a set of objects; 2. Strain or unit deformation
S	Section modulus
t	1. Time; 2. Thickness
T	1. Temperature; 2. Torsional moment
v	1. Velocity; 2. Unit shear stress (used in references not in this book)
V	1. Gross (total) shear force; 2. Vertical component of a force
w	1. Width; 2. Unit weight; 3. Unit of a uniformly distributed linear load (as on a beam)

W 1. Gross (total) value of a uniformly distributed load; 2. Gross (total) weight of an object

Greek Symbols

μ (mu) Coefficient of friction
ϕ (phi) Angle
Δ (delta) Deflection
θ (theta) Angle

1

Structural Wood

All wood materials used for construction receive some form of processing. This chapter deals with solid-sawn elements, which are used essentially in their natural state except for the sawing, surface finishing, and moisture curing that makes them presentable for use. In this case the source (tree) from which the wood comes is of prime concern, although many other factors must be considered.

1-1 Introduction

The particular type of tree from which wood comes is called the *species*. Although there are thousands of species of trees, most structural wood comes from a few dozen species that are selected for commercial lumber processing.

The two groups of trees used for building purposes are the *softwoods* and *hardwoods*. Softwoods like the pines and spruce are coniferous, or cone bearing, whereas hardwoods have broad leaves exemplified by the oaks and maples. The terms softwood and hardwood are not accurate indications of the degree of hardness of the various species of trees. Certain softwoods are as hard as the medium-density hardwoods, whereas some species of hardwoods have softer wood than some of the softwoods. The two species of trees used most extensively in the United States for structural members are Douglas fir and Southern pine, both of which are classified among the softwoods. Many other species, however, are also used for structural lumber. Although the terms *timber* and *structural lumber* are often used interchangeably, current usage tends to reserve *timber* for the structural wood members of larger cross-sectional area.

1-2 Tree Growth

The trees used for lumber in this country are exogenous; that is, they increase in size by a growth of new wood on the outer surface under the bark. The cross section of a tree trunk reveals the layers of new wood that are formed annually. These layers, called *annual rings,* are frequently composed of light and dark layers; the light ring, the springwood, is the wood grown in the spring of the year and the darker ring, is the summerwood. Thus the number of annual rings at the base of a tree indicates the age of the tree. The band of annual rings at the outer edge of the trunk is known as the *sapwood.* This band is frequently light colored. It contains the living cells and carries the sap from the roots to the leaves. As the tree ages, the sapwood gradually changes to *heartwood* and new sapwood is formed. The heartwood is usually darker in color than the sapwood. It is composed of the inactive cells and constitutes the major portion of the tree trunk. In general, the sapwood is light and more porous than the heartwood. The heartwood is denser and gives strength to the tree trunk. It is stronger and more durable than sapwood, but, if the wood is to be treated with a preservative, sapwood is desirable because of its absorptiveness.

The structure of trees consists of longitudinal bundles of wood fibers or cells. These small hollow fibers vary in shape and arrangement, affecting both appearance and physical properties of the various species. Smaller bands of fibers, called *medullary* or *wood rays,* radiate from the center of the tree trunk and serve to bind the structure together. The medullary rays are not distinctive in some species of trees but are pronounced in others; quarter-sawed oak, for example, shows these wood rays quite clearly.

1-3 Density of Wood

The difference in arrangement and size of the cell cavities and the thickness of the cell walls determine the specific gravity of various species of wood. The strength of wood is closely related to its density. The term *close grained* refers to wood with narrow, closely spaced annual rings. Certain woods, such as Douglas fir and Southern yellow pine, show a distinct contrast between springwood and summerwood, and the proportion of summerwood affords a visual basis for approximating strength and density. The weight of wood substance of all species is about 1.53 times the weight of water, but the wood cells contain air in varying degrees; hence the weights of species vary not only because of their density but also because of the moisture content. For purposes of computation in this book, the average weight of wood is taken as 35 lb per cu ft.

1-4 Defects in Lumber

Any irregularity in wood that affects its strength or durability is called a *defect*. Because of the natural characteristics of the material, several common defects are inherent in wood. The most common are described here.

A *knot* is a portion of a branch or limb that has been surrounded by subsequent growth of the tree. There are several types and classifications of knots, and the strength of a structural member is affected by the size and location of those it may contain. The grading rules for structural lumber are specific concerning the number, sizes, and position of knots, and their presence is considered when establishing the allowable unit stresses known as *design values*.

A *shake* is a separation along the grain, principally between the annual rings. The cross section of a shake is shown in Fig. 1-1*a*. Shakes reduce the resistance to shear, and consequently members subjected to bending are directly affected by their presence. The strength of members in longitudinal compression (columns, posts, etc.) is not greatly affected by shakes.

A *check* is a separation along the grain, the greater part of which occurs *across* the annual rings (Fig. 1-1*b*). Checks gener-

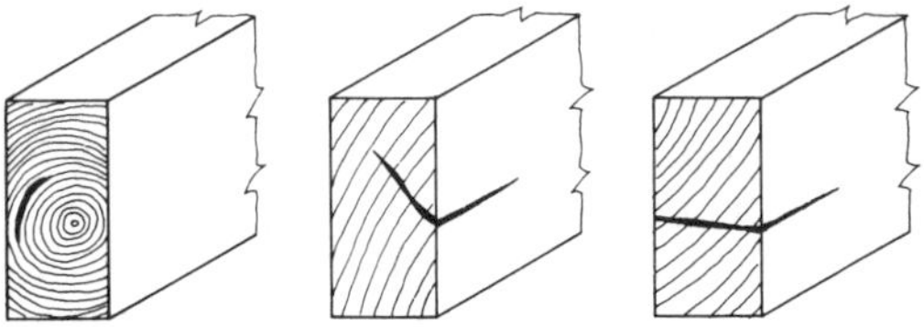

FIGURE 1-1. Defects in structural lumber.

ally arise from the process of seasoning. Like shakes, checks also reduce the resistance to shear.

A *split* is shown in Fig. 1-1*c*. It is defined as a lengthwise separation of the wood that extends through the piece from one surface to another.

Decay is the disintegration of wood substance due to the action of wood-destroying fungi. Decay is easily recognized, for the wood becomes soft, spongy, or crumbly. The growth of fungi is encouraged by air, moisture, and a favorable temperature. If air is excluded, for instance, when wood is constantly submerged, fungi cannot exist. Wood is often impregnated with preservatives such as coal tar and creosote to prevent growth of fungi. The development of fungi is also prevented by the application of paint to the wood when it is dry. The extent of decay is generally difficult to determine: therefore any form of decay is usually prohibited in structural grades of wood.

A *pitch pocket* is an opening parallel to the annual rings that contains pitch, either solid or liquid.

1-5 Seasoning of Wood

All green wood contains moisture, and the serviceability of wood is improved by its removal. The process of removing moisture from green wood is known as *seasoning;* it is accomplished by exposing lumber to the air for an extended period or by heating it in kilns. Whether *air dried* or *kiln dried,* seasoned wood is stiffer, stronger, and more durable than green wood. The removal of moisture results in the shrinkage of the fiber cells; side fiber walls shrink more than end walls and sapwood more than heartwood. This shrinkage causes internal stresses that result in checking and warping, but development of these defects can be controlled to some extent by proper seasoning procedures. The *moisture content* of wood is defined as the ratio of the weight of water in a specimen to the weight of the oven-dry wood expressed as a percentage.

1-6 Use Classification of Structural Lumber

Because the effects of natural defects on the strength of lumber vary with the type of loading to which an individual piece is subjected, structural lumber is classified according to its *size and use*. The four principal classifications are dimension, beams and stringers, posts and timbers, and decking. They are defined as follows:

Dimension. This consists of rectangular cross sections with nominal dimensions, 2 in. to 4 in. thick and 2 in. or more wide. This classification is further divided into *light framing* grades 2 in. to 4 in. wide, and *joists and planks* 5 in. and wider.

Beams and Stringers. Rectangular cross sections 5 in. or more thick and a width more than 2 in. greater than the thickness are graded for strength in bending when loaded on the narrow face.

Posts and Timbers. Square or nearly square cross sections with nominal dimensions 5 in. by 5 in. and larger are graded primarily for use as posts or columns but adapted to other uses where bending strength is not especially important.

Decking. This consists of lumber 2 in. to 4 in. thick, 6 in. and wider, with tongue and groove edges or grooved for spline on the narrow face. Decking is graded for use with the wide face placed flatwise in contact with supporting members.

There is some confusion in the terms used to refer to the dimensions of a rectangular cross section of wood. In the use classifications just described the term *thickness* is used for the smaller dimension and *width* is used for the larger dimension of an oblong section. However, when referring to beam sections (for beams that are vertically loaded), it is common to use width for the horizontal dimension (usually the smaller dimension) and depth for the vertical dimension (usually the larger dimension). We shall try not to compound this confusion, but unfortunately it already exists in the reference literature.

1-7 Nominal and Dressed Sizes

An individual piece of structural lumber is designated by its *nominal* cross-sectional dimensions. As an example, we speak of a 6 × 12″ (written 6 × 12), by which we mean a timber with a width of 6 in. and a depth of 12 in.; the length is variable. However, after being *dressed* or *surfaced* on four sides (S4S) the actual dimensions of this piece are $5\frac{1}{2} \times 11\frac{1}{2}$ in. The first two columns of Table 4-1 list nominal and standard dressed sizes of structural lumber in accordance with American Softwood Lumber Standard, PS 20–70, promulgated by the U.S. Department of Commerce.

Lumber is sold on the basis of the contents of the nominal size expressed in terms of *board feet*. A board foot is the content of a volume 12 × 12 × 1 in., 144 cu in., or $\frac{1}{12}$ cu ft.

1-8 Grading of Structural Lumber

Grading is necessary to identify the quality of lumber. Structural grades are established in relation to strength properties and use classification so that allowable stresses for design can be assigned (Chapter 3). Individual grades of the various species are given a commercial designation such as No. 1, No. 2, Select Structural, and Dense No. 2 by the grading-rules agency concerned. Some of these industry associations are the Southern Pine Inspection Bureau, the West Coast Lumber Inspection Bureau, and the Western Wood Products Association.

2

Unit Stresses

2-1 General

As a prelude to discussing the strength and behavior of structural wood under load, it is necessary to establish a clear understanding of the concept of *unit stress*. Throughout this book there will be many technical terms which may or may not be familiar, depending on the extent of the reader's experience in structural work. For those who have had some preparation in structural mechanics (statics and strength of materials) the material in this chapter will serve as a review. In any event, it is important that these terms and the concepts to which they apply be understood precisely.

2-2 Forces and Loads

A *force* is defined in mechanics as that which tends to change the state of rest or motion of a body. It may be considered as pushing or pulling a body at a definite point and in a definite direction. Such a force tends to give motion to a body at rest, but this tendency may be neutralized by the action of another force or forces. In building construction we are concerned primarily with forces in equilibrium; that is, with bodies at rest. The units of force are pounds, kilograms, tons, and so on, and in engineering practice the term *kip,* meaning 1000 pounds, is also widely used. The *weight* of a body is a vertical force due to gravity.

A *load* is the magnitude of pressure or tension due to a superimposed weight. The two types most common in engineering problems are *concentrated loads* and *uniformly distributed loads.*

A uniformly distributed load is a load of uniform magnitude per unit of length that extends over a portion of a member or over its entire length. A floor joist that supports board flooring is an example of a member supporting a uniformly distributed load. It should be noted that the weight of the joist itself also constitutes a uniformly distributed load.

A concentrated load is one that extends over so small a portion of the length of a beam or girder that it may be assumed to act at a point. A girder in a building receives concentrated loads at the points at which floor beams frame into it.

The term *dead load* is applied to the weight of the materials of construction; that is, to the weight of beams, girders, flooring, partitions, and so on. The *live load* represents the probable load due to occupancy of a building and includes the weight of human occupants, furniture, equipment, stored materials, and snow. The sum of all dead and live loads is the *total load.*

Figure 2-1*a* represents the floor framing of a bay in a building framed with structural timber. The columns are spaced 14 ft on centers in one direction and 16 ft in the other. The girders span the shorter direction and the beams the longer. Each girder supports a beam at the center of its 14 ft span. The beams in turn support plank flooring (not shown in the drawing) spanning parallel to the girders. Let us assume that the total load on the floor, live and dead, is 80 pounds per square floor (psf). The area of floor supported by the center beam in this bay is shown by hatching in the figure; it is equal to 7×16, or 112 sq ft. Therefore the total uniformly distributed load on the beam is $112 \times 80 = 8960$ lb. Figure 2-1*b* is the conventional diagram that represents this beam and loading. In this text we use W to represent the total uniformly distributed load and w, the uniformly distributed load per linear foot. In this instance $W = 8960$ lb and $w = 8960 \div 16 = 560$ lb per lin ft (lb/ft).

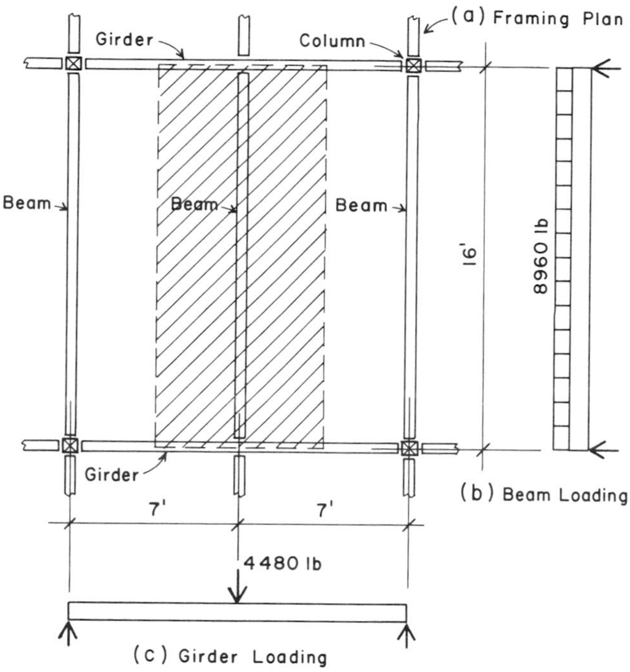

FIGURE 2-1.

The loading of this beam is symmetrical about the center of its span (Fig. 2-1*b*) and consequently it exerts a force of 8960 ÷ 2 = 4480 lb at each end on the girders. This force becomes a concentrated load on the girders as indicated in Fig. 2-1*c*. If this were a problem involving the design of the girders, we would have to take into account the loads transmitted by beams in adjacent bays. Assuming that adjacent bays have the same layout and span lengths as this one, the concentrated load at midspan of the girders would be 4480 × 2 = 8960 lb.

2-3 Direct Stress

A *stress* in a body is an internal resistance to an external force. When the external force acts along the axis of the body, it is called an *axial force* or *axial load*. In Fig. 2-2 the short wood post

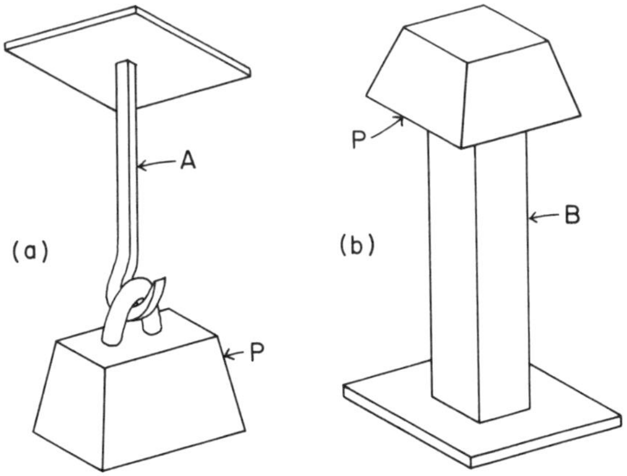

FIGURE 2-2. Direct stress.

B supports an axial load caused by the weight *P*. The load exerts a compressive force on the post, tending to shorten it, and this tendency is resisted by the compressive stress developed in the post. The stress produced in the post under this condition of axial loading is called *direct stress*.

A characteristic of direct stress is that the internal resistance may be assumed to be evenly distributed over the cross-sectional area of the body under stress. Thus, if *P* in Fig. 2-2 is 6400 lb and *B* is a nominal 6 × 6 post (dressed area = 30.25 sq in.), each square inch of the post cross section is stressed to 6400 ÷ 30.25 = 212 pounds per square inch (psi). This stress per unit area is called the *unit stress* to distinguish it from the internal force of 6400 lb. By calling the load or external force *P*, the area of cross section *A*, and the unit stress *f*, this fundamental relationship governing direct stress may be stated by the following equation:

$$f = \frac{P}{A} \quad \text{or} \quad P = fA \quad \text{or} \quad A = \frac{P}{f}$$

When using this equation, remember the two assumptions on which it is based: the loading is axial and the stresses are evenly (uniformly) distributed over the cross section. Note also that if any two of the quantities are known the third may be found.

2-4 Kinds of Stress

The three basic kinds of stress with which we are principally concerned are *compression, tension,* and *shear*. As we observed in Sec. 2-3, compressive stress results from a force that tends to compress or crush a member. The stress of 212 psi found in that section is a *compressive stress*.

A *tensile stress* is the stress that results from a force that tends to stretch or elongate a member (Fig. 2-1*a*). The lower chord and certain web members of trusses and trussed rafters (Chapter 15) are in tension. If the total axial tensile force (total stress) in a member is known as well as its area of cross section, the unit tensile stress may be found from the basic direct stress formula $f = P/A$.

A *shearing stress* results from the tendency of two equal and parallel forces, acting in opposite directions, to cause adjoining surfaces of a member to slide one on the other. Figure 2-3*a* represents a beam with a uniformly distributed load. There is a tendency for the beam to fail by dropping down between the supports as indicated in Fig. 2-3*b*. This is an example of *vertical shear*. Figure 2-3*c* shows an exaggerated bending action in a beam and the failure by portions of the beam sliding horizontally. Figure 2-3*d* illustrates the tendency of the lower chord of a roof truss to fail by shearing action induced by the thrust of the upper chord at an end joint. Both Figs. 2-3*c* and *d* are illustrations of *horizontal shear*. It is shown later that shear failures in wood beams are due to horizontal, not vertical, shear. This is true because the shearing resistance of wood is much less parallel to the grain than it is across the grain. The method of determining maximum horizontal unit shearing stress in beams is explained in Chapter 6.

In addition to shear, *bending stresses* (both compression and tension) are developed in beams under load. The controlling

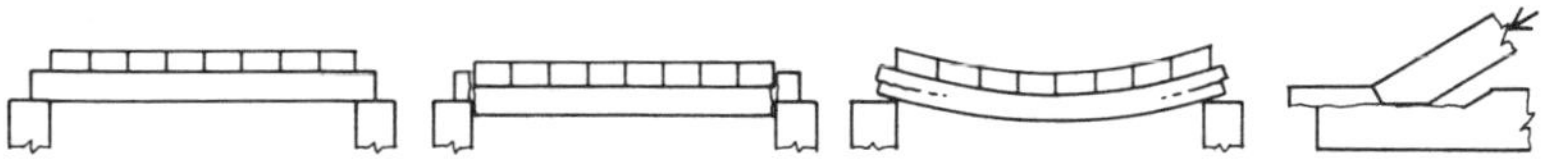

(a) (b) (c) (d)

FIGURE 2-3. Development of shear stress.

bending stress, called *extreme fiber stress,* is discussed in Chapter 7.

2-5 Deformation

Whenever a body is subjected to a force, there is a change in its size or shape; this change is called *deformation*. Regardless of the magnitude of the force, some deformation always takes place, although often it is so small that measurement is difficult, even with the most sensitive instruments. Under axial compression and tension the deformations are a shortening and a lengthening, respectively. When a force acts on a member in a manner that causes bending (such as a load on a beam), the deformation is called *deflection*. Computation of the deflection of wood beams under different loading conditions is discussed in Chapter 8.

2-6 Elastic Limit

Current design practice for structural wood members is based on elastic theory, which postulates that *deformations are directly proportional to stresses*. In other words, if an applied force (as measured by its resulting unit stress) produces a certain deformation, twice the force will produce twice the amount of deformation. This relationship between stress and deformation holds true only up to a certain limit, after which the deformation begins to increase at a faster rate than the increments of the applied load. The unit stress at which this occurs is called the *elastic limit* or the *proportional limit* of a material.

Elasticity is the property of a material that enables it to return to its original size and shape when the load to which it has been subjected is removed. This occurs, however, *only when the unit stress does not exceed the elastic limit*. Beyond the elastic limit a permanent deformation, called a *permanent set,* remains in the member. The allowable unit stresses used in the design of wood structural members are established so that the elastic or proportional limit of the material will not be exceeded under service loads.

2-7 Ultimate Strength

The *ultimate strength* of a material is defined as the unit stress that occurs at or just before rupture. Some structural materials possess considerable reserve strength between the elastic limit and the ultimate strength, but this "inelastic" strength is not taken into account directly under the elastic theory of structural design.

Note that strength properties for various species of wood are not so clearly defined as they are for some other construction materials such as structural steel. Tests of specimens of the same species and size, and in the same condition, may exhibit a considerable spread in strength values. This variability in test results is, of course, taken into account when allowable stresses (design values) for the different species and grades of structural lumber are established.

2-8 Modulus of Elasticity

The *modulus of elasticity* of a material is a measure of its *stiffness*. A specimen of steel deforms a certain amount when subjected to a given load, but a wood specimen of the same dimensions subjected to the same load deforms probably 15 to 20 times as much. We say the steel is *stiffer* than the wood. The ratio between the unit stress and the unit deformation, provided the unit stress does not exceed the elastic limit of the material, is called the modulus of elasticity of the material. It is denoted by the symbol E and is expressed in pounds per square inch. For structural steel $E = 29{,}000{,}000$ psi [200 GPa] and for wood, depending on the species and grade, it varies from something less than 1,000,000 psi to about 1,900,000 psi [7–13 GPa]. The modulus of elasticity of structural lumber is used in computing the deflection of beams.

2-9 Allowable Unit Stress

An *allowable unit stress* is the stress used in design computations and represents the maximum unit stress of a particular kind considered acceptable in a structural member subjected to loads. Allowable unit stresses are sometimes known as *working stresses* and are called *design values* in the 1986 edition of the *National Design Specification for Wood Construction* issued by the National Forest Products Association (Ref. 1); from now it is referred to as the NDS. The procedures for establishing allowable unit stresses in tension, compression, shear, and bending are different for different materials and are prescribed in specifications promulgated by the American Society for Testing and Materials. Allowable unit stresses for structural lumber are discussed in Chapter 3.

3

Design Values

3-1 General

There are many factors to be considered in determining the allowable unit stresses for structural lumber. Numerous tests by the Forest Products Laboratory of the U.S. Department of Agriculture made on material free from defects have resulted in a tabulation known as *clear wood strength values*. To obtain design values the clear wood values are reduced by factors that take into consideration the loss of strength from defects, size and position of knots, size of member, degree of density, and condition of seasoning. These modifications are made in accordance with the provisions of ASTM Designation D-245, "Methods for Establishing Structural Grades and Related Allowable Properties for Visually Graded Lumber."

3-2 Design Value Tables

The general reference for allowable stresses and modulus of elasticity values to be used for design work is the publication that is prepared as a supplement to the NDS entitled *Design Values for Wood Construction*. Table 3-1 is adapted from this publication

TABLE 3-1. Design Values for Structural Lumber[a]

Species and Commercial Grade	Size Classification	Design Values in Psi: Extreme Fiber in Bending, F_b: Single Member Uses	Repetitive Member Uses	Tension Parallel to Grain F_t	Horizontal Shear F_v	Compression Perpendicular to Grain $F_{c\perp}$	Compression Parallel to Grain F_c	Modulus of Elasticity E
Douglas Fir–Larch (Surfaced Dry or Surfaced Green, Used at 19% Maximum Water Content)								
Dense Select Structural	2–4 in. thick, 2–4 in. wide	2450	2800	1400	95	730	1850	1,900,000
Select Structural		2100	2400	1200	95	625	1600	1,800,000
Dense No. 1		2050	2400	1200	95	730	1450	1,900,000
No. 1		1750	2050	1050	95	625	1250	1,800,000
Dense No. 2		1700	1950	1000	95	730	1150	1,700,000
No. 2		1450	1650	850	95	625	1000	1,700,000
No. 3		800	925	475	95	625	600	1,500,000
Appearance		1750	2050	1050	95	625	1500	1,800,000
Stud		800	925	475	95	625	600	1,500,000
Construction	2–4 in. thick, 4 in. wide	1050	1200	625	95	625	1150	1,500,000
Standard		600	675	350	95	625	925	1,500,000
Utility		275	325	175	95	625	600	1,500,000

Dense Select Structural	2–4 in. thick, 5 in. and wider	2100	2400	1400	95	730	1650	1,900,000
Select Structural		1800	2050	1200	95	625	1400	1,800,000
Dense No. 1		1800	2050	1200	95	730	1450	1,900,000
No. 1		1500	1750	1000	95	625	1250	1,800,000
Dense No. 2		1450	1700	775	95	730	1250	1,700,000
No. 2		1250	1450	650	95	625	1050	1,700,000
No. 3		725	850	375	95	625	675	1,500,000
Appearance		1500	1750	1000	95	625	1500	1,800,000
Stud		725	850	375	95	625	675	1,500,000
Dense Select Structural	Beams and stringers	1900	—	1100	85	730	1300	1,700,000
Select Structural		1600	—	950	85	625	1100	1,600,000
Dense No. 1		1550	—	775	85	730	1100	1,700,000
No. 1		1300	—	675	85	625	925	1,600,000
No. 2		875	—	425	85	625	600	1,300,000
Dense Select Structural	Posts and timbers	1750	—	1150	85	730	1350	1,700,000
Select Structural		1500	—	1000	85	625	1150	1,600,000
Dense No. 1		1400	—	950	85	730	1200	1,700,000
No. 1		1200	—	825	85	625	1000	1,600,000
No. 2		750	—	475	85	625	700	1,300,000
Select Dex	Decking	1750	2000	—	—	625	—	1,800,000
Commercial Dex		1450	1650	—	—	625	—	1,700,000

TABLE 3-1. Design Values for Structural Lumber[a] (*continued*)

		Design Values in Psi						
		Extreme Fiber in Bending, F_b						
Species and Commercial Grade	Size Classification	Single Member Uses	Repetitive Member Uses	Tension Parallel to Grain F_t	Horizontal Shear F_v	Compression Perpendicular to Grain $F_{c\perp}$	Compression Parallel to Grain F_c	Modulus of Elasticity E
Southern Pine (Surfaced Dry, Used at 19% Maximum Moisture Content)								
Select Structural	2–4 in. thick, 2–4 in. wide	2000	2300	1150	100	565	1550	1,700,000
Dense Select Structural		2350	2700	1350	100	660	1800	1,800,000
No. 1		1700	1950	1000	100	565	1250	1,700,000
No. 1 Dense		2000	2300	1150	100	660	1450	1,800,000
No. 2		1400	1650	825	90	565	975	1,600,000
No. 2 Dense		1650	1900	975	90	660	1150	1,600,000
No. 3		775	900	450	90	565	575	1,400,000
No. 3 Dense		925	1050	525	90	660	675	1,500,000
Stud		775	900	450	90	565	575	1,400,000
Construction	2–4 in. thick, 4 in. wide	1000	1150	600	100	565	1100	1,400,000
Standard		575	675	350	90	565	900	1,400,000
Utility		275	300	150	90	565	575	1,400,000
Select Structural	2–4 in. thick, 5 in. and wider	1750	2000	1150	90	565	1350	1,700,000

Dense Select Structural		2050	2350	1300	90	660	1600	1,800,000
No. 1		1450	1700	975	90	565	1250	1,700,000
No. 1 Dense		1700	2000	1150	90	660	1450	1,800,000
No. 2		1200	1400	675	90	565	1000	1,600,000
No. 2 Dense		1400	1650	725	90	660	1200	1,600,000
No. 3		700	800	350	90	565	625	1,400,000
No. 3 Dense		825	925	425	90	660	725	1,500,000
Stud		725	850	350	90	565	625	1,400,000
Dense Standard Decking	2–4 in. thick, 2 in. and wider	2000	2300	—	—	660	—	1,800,000
Select Decking		1400	1650	—	—	565	—	1,600,000
Dense Select Decking	Decking	1650	1900	—	—	660	—	1,600,000
Commercial Decking		1400	1650	—	—	565	—	1,600,000
Dense commercial decking		1650	1900	—	—	660	—	1,600,000
Dense Structural 86	2–4 in. thick	2600	3000	1750	155	660	2000	1,800,000
Dense Structural 72		2200	2550	1450	130	660	1650	1,800,000
Dense Structural 65		2000	2300	1300	115	660	1500	1,800,000

TABLE 3-1. Design Values for Structural Lumber[a] (*continued*)

		Design Values in Psi						
		Extreme Fiber in Bending, F_b						
Species and Commercial Grade	Size Classification	Single Member Uses	Repetitive Member Uses	Tension Parallel to Grain F_t	Horizontal Shear F_v	Compression Perpendicular to Grain $F_{c\perp}$	Compression Parallel to Grain F_c	Modulus of Elasticity E
Southern Pine (Surfaced Green, Used Any Condition)								
No. 1 SR	5 in. and thicker	1350	—	875	110	375	775	1,500,000
No. 1 Dense SR		1550	—	1050	110	440	925	1,600,000
No. 2 SR		1100	—	725	95	375	625	1,400,000
No. 2 Dense SR		1250	—	850	95	440	725	1,400,000
Dense Structural 86	$2\frac{1}{2}$ in. and thicker	2100	2400	1400	145	440	1300	1,600,000
Dense Structural 72		1750	2050	1200	120	440	1100	1,600,000
Dense Structural 65		1600	1800	1050	110	440	1000	1,600,000

[a] Values listed are for normal duration loading. Data adapted from *National Design Specifications for Wood Construction*, 1986 edition (Ref. 1), with permission of the publisher, National Forest Products Association. The table in the reference document gives values for several additional wood species and has extensive footnotes.

and presents values for two popular species of structural lumber: Douglas fir–larch and Southern pine. To obtain values from the table the following data must be determined for a particular piece of lumber.

1. *Species*. The NDS publication lists values for more than 40 different species, only two of which are included in Table 3-1.
2. *Moisture Condition at Time of Use*. The moisture condition assumed for the table values is given with the species designation in the table. Adjustments for other moisture conditions are described in the footnotes or in the various specifications in the NDS.
3. *Grade*. This is indicated in the first column of the table which is based on visual grading standards.
4. *Size or Use*. The second column of the table identifies size ranges or usage (e.g., beams and stringers) of the lumber. Note that for Douglas fir–larch the grade "select structural" appears six times for various sizes and uses.
5. *Structural Function*. Values are given for stresses in flexure, tension, shear, and compression, and for the modulus of elasticity.

In the reference document there are 20 footnotes to the table, extending over two full pages. Data from Table 3-1 will be used in various design examples in this book and some of the issues treated in the document footnotes will be explained. In many situations there are modifications to the design values, as described in the next section.

3-3 Modifications of Design Values

The values given in Table 3-1 are basic references for establishing the allowable values to be used for design. These values are based on some defined norms, and in many situations the design values will be modified for actual use in structural computations. In some cases the form of the modification is a simple increase or decrease that is achieved by a percentage increase or reduction factor. In other situations the modification is more complex, such as the modification of allowable compression for slenderness effects. Some of the modifications are described in the footnotes to the table from which Table 3-1 is adapted. Other modifications are described in Chapter 2 or in the various sections of the NDS that deal with specific types of problems. The following are some of the major types of modifications.

Moisture. The table gives the assumed moisture condition on which the table values are based. Increases may be permitted in some values for wood that is cured to a lower moisture value. If exposed to weather, a wet usage condition may require some reductions of values.

Load Duration. The table values are based on so-called *normal* duration load, which is actually rather meaningless. Increases in the design values are permitted for various degrees of short duration loading. A decrease in some values may be required for loading that is prolonged over the life of the structure (basically referring to dead load). Table 3-2 presents a summary of the NDS requirements for adjustment for load duration.

TABLE 3-2. Modification of Design Values for Load Duration[a]

Duration of Load and General Use	Multiply Design Values By
Ten years or more at the full load limit of a member (as for members carrying only dead load, such as headers in walls)	0.90
Two months duration, as for snow	1.15
Seven days duration (as for roof loads where no snow pack occurs)	1.25
Maximum force of wind or earthquake	1.33
Impact (such as wheel bumps, sudden braking of moving equipment, and slamming of heavy doors)	2.00

[a] Adapted from specifications in *National Design Specifications for Wood Construction*, 1986 edition, with permission of the publishers, National Forest Products Association.

Temperature. Where prolonged exposure to temperatures over 150°F exist, some values must be reduced.

Treatments. Impregnation with chemicals for enhanced resistance to rot, vermin, and insects, or fire may require reductions in some values.

Size. Effectiveness in flexure is reduced in beams of depths exceeding 12 in., as described in Sec. 7-4.

Slenderness. Various modifications may be required for beams or columns with a tendency to fail in buckling.

Load Orientation to Grain. The table gives separate values for allowable compression with respect to the grain direction in the wood. In some situations the load direction may be other than parallel (0°) or perpendicular (90°) to the grain and a specific value for stress must be derived, as described in Sec. 3-4.

Specific usage conditions must be carefully studied to ascertain the amount of modification required for a given design situation.

3-4 Modifications for Loading at Angles to the Grain

Under the condition shown in Fig. 3-1 the load from member *B* exerts a compressive stress on member *A* on a surface inclined to the grain. The compressive strength of wood is greatest parallel to the grain and is least perpendicular to the grain. The allowable unit compressive stress on an inclined surface is determined from the following expression, known as the Hankinson formula:

$$F_n = \frac{F_c \times F_{c\perp}}{F_c \times \sin^2 \theta + F_{c\perp} \times \cos^2 \theta}$$

in which F_n = allowable unit stress acting perpendicular to the inclined surface,

F_c = allowable unit stress in compression parallel to the grain,

$F_{c\perp}$ = allowable unit stress in compression perpendicular to the grain,

θ = angle between the direction of the load and the direction of the grain.

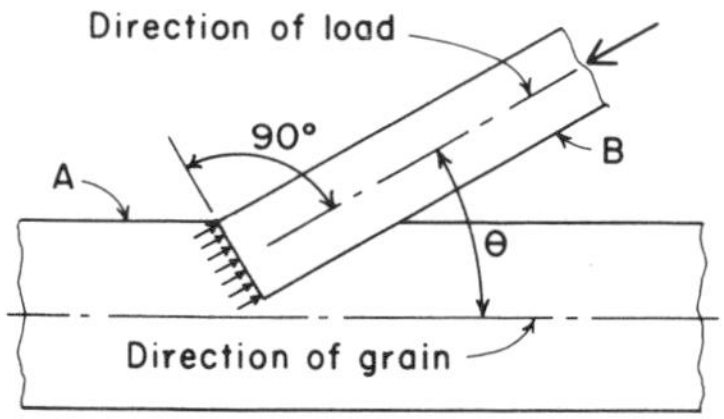

FIGURE 3-1.

When the load is applied parallel to the grain, θ is zero. When the load is applied perpendicular to the grain, θ is 90°. Table 3-3 gives values of $\sin^2 \theta$ and $\cos^2 \theta$ for various values of θ (theta).

Example. Two timbers 6 in. wide consisting of Southern pine, No. 1 Dense SR grade, are framed together as indicated in Fig. 3-1. The angle between the two pieces is 30°. Compute the allowable unit stress on the inclined bearing surface.

Solution: Referring to Table 3-1, we find that the allowable unit compressive stress parallel to grain is $F_c = 925$ psi, and perpendicular to grain, $F_{c\perp} = 440$ psi. Values of $\sin^2 \theta$ and $\cos^2 \theta$, when $\theta = 30°$, are taken from Table 3-3. Then, substituting in the Hankinson formula the allowable unit compressive stress on the

TABLE 3-3 Values for Use in the Hankinson Formula (see Fig. 3-1)

$\sin^2 \theta$	θ (degrees)	$\cos^2 \theta$	$\sin^2 \theta$	θ (degrees)	$\cos^2 \theta$
0.00000	0	1.00000	0.50000	45	0.50000
0.00760	5	0.99240	0.58682	50	0.41318
0.03015	10	0.96985	0.67101	55	0.32899
0.06698	15	0.93302	0.75000	60	0.25000
0.11698	20	0.88302	0.82140	65	0.17860
0.17860	25	0.82140	0.88302	70	0.11698
0.25000	30	0.75000	0.93302	75	0.06698
0.32899	35	0.67101	0.96985	80	0.03015
0.41318	40	0.58682	0.99240	85	0.00760
0.50000	45	0.50000	1.00000	90	0.00000

inclined surface is

$$F_n = \frac{925 \times 440}{(925 \times 0.25) + (440 \times 0.75)} = 725 \text{ psi}$$

A similar modification is required with the use of some fasteners, such as bolts and split-ring connectors. As an alternative to the computations just illustrated, the functions of the Hankinson formula may be displayed in a graphical form and necessary modifications can be approximated from the graph. The use of such a method is illustrated in Sec. 13-2.

Problem 3-4-A. Two Douglas fir–larch timbers of Select Structural grade frame together as shown in Fig. 3-1. The angle between the members is 45°. Determine the allowable unit compressive stress on the inclined contact surface.

4

Properties of Sections

4-1 General

In addition to the strength of wood, signified by allowable unit stresses, the performance of a structural member also depends on the size and shape of its cross section. These two factors are taken into account by the *properties* of the section; they are independent of the material of which the member is made. In this chapter we consider the definition and nature of a few of these properties to serve as background for their later application in the design of structural members.

4-2 Centroids

The *center of gravity* of a solid is an imaginary point at which all its weight may be considered to be concentrated or the point through which the resultant weight passes. The point in a plane area that corresponds to the center of gravity of a very thin plate of the same area and shape is called the *centroid* of the area.

When a beam spanning between two walls tends to bend because of an applied load, the fibers above a certain plane in the beam are in compression and those below the plane are in ten-

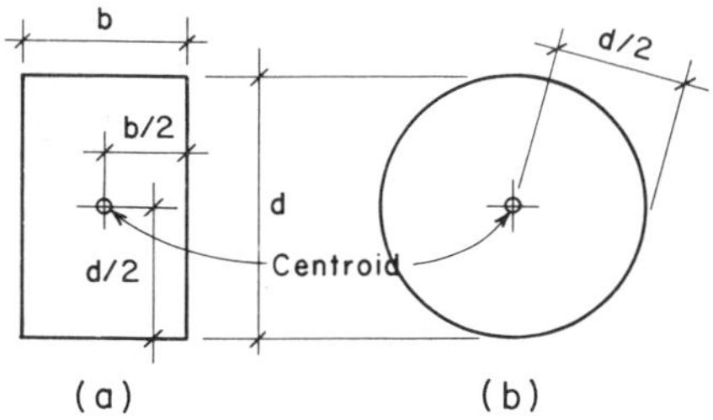

FIGURE 4-1.

sion. This plane is called the *neutral surface*. The line in which the neutral surface cuts the cross section of the beam is called the *neutral axis*. The neutral axis passes through the centroid of the section; thus it is important that we know the position of the centroid.

The position of the centroid for symmetrical sections is readily determined. If the section possesses a line of symmetry, the centroid will obviously be on that line; if there are two lines of symmetry, the centroid will be at their point of intersection; for example, the rectangular beam cross section shown in Fig. 4-1*a* has its centroid at its geometrical center, the point of intersection of the diagonals. The centroid of a circular (pole) cross section is at its center (Fig. 4-1*b*).

With respect to dimensional notation, the letter b generally represents the breadth of the face of a member on which the load is applied. The letter d represents the depth or height of the beam face parallel to the direction of the line of action of the load. Sometimes the depth is represented by the letter h, but we follow the more general structural design practice and let d denote the depth of a beam cross section.

4-3 Moment of Inertia

Figure 4-2*a* indicates a rectangular section of breadth b and depth d with horizontal axis X–X passing through its centroid at distance $c = d/2$ from the top. Within the section a represents an infinitely small area at z distance from axis X–X. If we multiply this infinitesimal area by the square of its distance to the axis, we

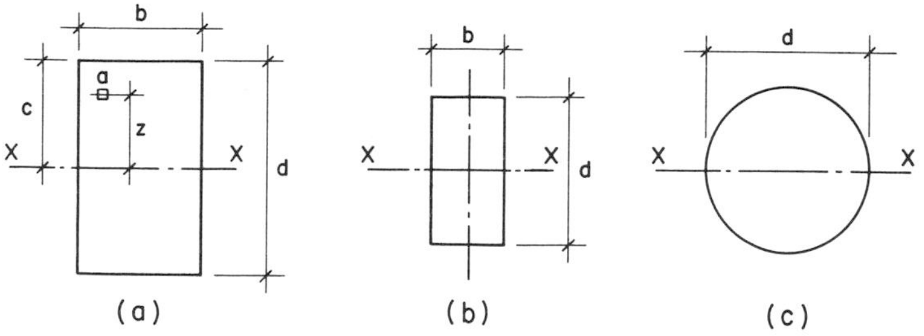

FIGURE 4-2.

have the quantity ($a \times z^2$). The entire area of the section is made up of an infinite number of these small elementary areas at various distances above and below axis X–X. If we use the Greek letter Σ to represent the sum of an infinite number, we may write Σaz^2, which means the sum of all the infinitely small areas (of which the section is composed), multiplied by the square of their distances from the X–X axis. This quantity is called the *moment of inertia* of the section and is denoted by the letter I. More specifically, $I_{X\text{-}X} = \Sigma az^2$ is the moment of inertia with respect to the axis marked X–X.

We may now define moment of inertia as *the sum of the products obtained by multiplying all the infinitely small areas by the square of their distances from an axis*. The linear dimensions of cross sections of structural members are given in units of inches and, because the moment of inertia involves an area multiplied by the square of a distance, it is expressed in inches to the fourth power, written in.^4 The derivation of formulas for computing moments of inertia of various shapes is most readily accomplished by use of the calculus. It is beyond the scope of this book to derive such formulas, but the application of two of them is illustrated here.

Rectangles. Consider the rectangle shown in Fig. 4-2*b*. Its breadth is b and its depth is d. The two major axes are X–X and Y–Y, both of which pass through the centroid of the section. It can be shown that the moment of inertia of a rectangular section about an axis passing through the centroid and parallel to the base is $I_{X-X} = bd^3/12$. With respect to the vertical axis the expression would be $I_{Y-Y} = db^3/12$. In the design of wood beams and planks, however, it is customary to work with I_{X-X} only and to consider the top (loaded) face as b in the formula.

Example. Compute the moment of inertia of the cross section of a 6 × 12 timber (dressed dimensions 5.5 × 11.5 in. [140 × 290 mm]) with respect to a horizontal axis passing through the centroid and parallel to the shorter side.

Solution: In referring to Fig. 4.2*b*, width b = 5.5 in. [140 mm] and depth d = 11.5 in. [290 mm]. Then

$$I_{X-X} = \frac{bd^3}{12} = \frac{(5.5)(11.5)^3}{12} = 697 \text{ in.}^4 \ [285 \times 10^6 \text{ mm}^4]$$

Table 4-1 gives the moment of inertia for several standard dressed sizes of structural lumber and thus it is unnecessary to solve this formula. In referring to the table, enter on the line for nominal size 6 × 12 and read 697.068 in the fourth column. Although the table data give the numbers to the third decimal place, the first three digits are usually all the accuracy required for most structural computations.

If this 6 × 12 timber is used with the 12 in. side flat, I_{X-X} is computed using b = 11.5 and d = 5.5, yielding the value given in the table for the nominal size 12 × 6. The reference diagram at the top of the table indicates the constant use of the dimensions designated b and d.

Circular Sections. The moment of inertia of a circular cross section, such as that of a pole or a foundation pile, is the same about *any* axis passing through its centroid. The formula for this condition is $I_{X-X} = \pi d^4/64$. Because axis X–X in Fig. 4-2*c* may be any axis through its centroid, it is customary to use the symbol I_0. Then, if the actual diameter of the member is 10 in. [250 mm],

$$I_0 = \frac{\pi d^4}{64} = \frac{3.1416(10)^4}{64} = 491 \text{ in.}^4 \ [192 \times 10^6 \text{ mm}^4]$$

TABLE 4-1. Properties of Structural Lumber of Standard Dressed Sizes (S4S)

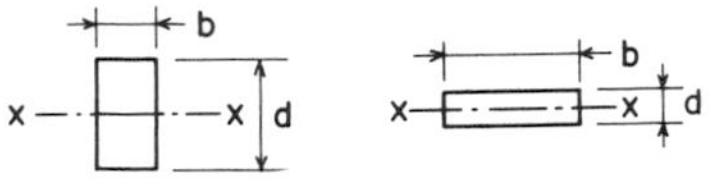

Nominal size b(inches)d	Standard dressed size (S4S) b(inches)d	Area of Section A	Moment of inertia I	Section modulus S	Approx. weight*
2 x 3	1-1/2 x 2-1/2	3.750	1.953	1.563	0.911
2 x 4	1-1/2 x 3-1/2	5.250	5.359	3.063	1.276
2 x 5	1-1/2 x 4-1/2	6.750	11.391	5.063	1.641
2 x 6	1-1/2 x 5-1/2	8.250	20.797	7.563	2.005
2 x 8	1-1/2 x 7-1/4	10.875	47.635	13.141	2.643
2 x 10	1-1/2 x 9-1/4	13.875	98.932	21.391	3.372
2 x 12	1-1/2 x 11-1/4	16.875	177.979	31.641	4.102
2 x 14	1-1/2 x 13-1/4	19.875	290.775	43.891	4.831
3 x 1	2-1/2 x 3/4	1.875	0.088	0.234	0.456
3 x 2	2-1/2 x 1-1/2	3.750	0.703	0.938	0.911
3 x 4	2-1/2 x 3-1/2	8.750	8.932	5.104	2.127
3 x 5	2-1/2 x 4-1/2	11.250	18.984	8.438	2.734
3 x 6	2-1/2 x 5-1/2	13.750	34.661	12.604	3.342
3 x 8	2-1/2 x 7-1/4	18.125	79.391	21.901	4.405
3 x 10	2-1/2 x 9-1/4	23.125	164.886	35.651	5.621
3 x 12	2-1/2 x 11-1/4	28.125	296.631	52.734	6.836
3 x 14	2-1/2 x 13-1/4	33.125	484.625	73.151	8.051
3 x 16	2-1/2 x 15-1/4	38.125	738.870	96.901	9.266
4 x 1	3-1/2 x 3/4	2.625	0.123	0.328	0.638
4 x 2	3-1/2 x 1-1/2	5.250	0.984	1.313	1.276
4 x 3	3-1/2 x 2-1/2	8.750	4.557	3.646	2.127
4 x 4	3-1/2 x 3-1/2	12.250	12.505	7.146	2.977
4 x 5	3-1/2 x 4-1/2	15.750	26.578	11.813	3.828
4 x 6	3-1/2 x 5-1/2	19.250	48.526	17.646	4.679
4 x 8	3-1/2 x 7-1/4	25.375	111.148	30.661	6.168
4 x 10	3-1/2 x 9-1/4	32.375	230.840	49.911	7.869
4 x 12	3-1/2 x 11-1/4	39.375	415.283	73.828	9.570
4 x 14	3-1/2 x 13-1/4	46.375	678.475	102.411	11.266
4 x 16	3-1/2 x 15-1/4	53.375	1034.418	135.66	12.975
5 x 2	4-1/2 x 1-1/2	6.750	1.266	1.688	1.641
5 x 3	4-1/2 x 2-1/2	11.250	5.859	4.688	2.734
5 x 4	4-1/2 x 3-1/2	15.750	16.078	9.188	3.828
5 x 5	4-1/2 x 4-1/2	20.250	34.172	15.188	4.922
6 x 1	5-1/2 x 3/4	4.125	0.193	0.516	1.003
6 x 2	5-1/2 x 1-1/2	8.250	1.547	2.063	2.005
6 x 3	5-1/2 x 2-1/2	13.750	7.161	5.729	3.342
6 x 4	5-1/2 x 3-1/2	19.250	19.651	11.229	4.679
6 x 6	5-1/2 x 5-1/2	30.250	76.255	27.729	7.352
6 x 8	5-1/2 x 7-1/2	41.250	193.359	51.563	10.026
6 x 10	5-1/2 x 9-1/2	52.250	392.963	82.729	12.700
6 x 12	5-1/2 x 11-1/2	63.250	697.068	121.229	15.373
6 x 14	5-1/2 x 13-1/2	74.250	1127.672	167.063	18.047
6 x 16	5-1/2 x 15-1/2	85.250	1706.776	220.229	20.720
6 x 18	5-1/2 x 17-1/2	96.250	2456.380	280.729	23.394
6 x 20	5-1/2 x 19-1/2	107.250	3398.484	348.563	26.068
6 x 22	5-1/2 x 21-1/2	118.250	4555.086	423.729	28.741
6 x 24	5-1/2 x 23-1/2	129.250	5948.191	506.229	31.415

* Weight in pounds per foot, based on average density of 35 pcf [560 kg/m^3].

Source: Compiled from data in the *National Design Specification for Wood Construction* (Ref. 1), with permission of the publishers, National Forest Products Association.

TABLE 4-1. *(Continued)*

Nominal size b(inches)d	Standard dressed size (S4S) b(inches)d	Area of Section A	Moment of inertia I	Section modulus S	Approx. weight*
8 x 1	7-1/4 x 3/4	5.438	0.255	0.680	1.322
8 x 2	7-1/4 x 1-1/2	10.875	2.039	2.719	2.643
8 x 3	7-1/4 x 2-1/2	18.125	9.440	7.552	4.405
8 x 4	7-1/4 x 3-1/2	25.375	25.904	14.803	6.168
8 x 6	7-1/2 x 5-1/2	41.250	103.984	37.813	10.026
8 x 8	7-1/2 x 7-1/2	56.250	263.672	70.313	13.672
8 x 10	7-1/2 x 9-1/2	71.250	535.859	112.813	17.318
8 x 12	7-1/2 x 11-1/2	86.250	950.547	165.313	20.964
8 x 14	7-1/2 x 13-1/2	101.250	1537.734	227.813	24.609
8 x 16	7-1/2 x 15-1/2	116.250	2327.422	300.313	28.255
8 x 18	7-1/2 x 17-1/2	131.250	3349.609	382.813	31.901
8 x 20	7-1/2 x 19-1/2	146.250	4634.297	475.313	35.547
8 x 22	7-1/2 x 21-1/2	161.250	6211.484	577.813	39.193
8 x 24	7-1/2 x 23-1/2	176.250	8111.172	690.313	42.839
10 x 1	9-1/4 x 3/4	6.938	0.325	0.867	1.686
10 x 2	9-1/4 x 1-1/2	13.875	2.602	3.469	3.372
10 x 3	9-1/4 x 2-1/2	23.125	12.044	9.635	5.621
10 x 4	9-1/4 x 3-1/2	32.375	33.049	18.885	7.869
10 x 6	9-1/2 x 5-1/2	52.250	131.714	47.896	12.700
10 x 8	9-1/2 x 7-1/2	71.250	333.984	89.063	17.318
10 x 10	9-1/2 x 9-1/2	90.250	678.755	142.896	21.936
10 x 12	9-1/2 x 11-1/2	109.250	1204.026	209.396	26.554
10 x 14	9-1/2 x 13-1/2	128.250	1947.797	288.563	31.172
10 x 16	9-1/2 x 15-1/2	147.250	2948.068	380.396	35.790
10 x 18	9-1/2 x 17-1/2	166.250	4242.836	484.896	40.408
10 x 20	9-1/2 x 19-1/2	185.250	5870.109	602.063	45.026
10 x 22	9-1/2 x 21-1/2	204.250	7867.879	731.896	49.644
10 x 24	9-1/2 x 23-1/2	223.250	10274.148	874.396	54.262
12 x 1	11-1/4 x 3/4	8.438	0.396	1.055	2.051
12 x 2	11-1/4 x 1-1/2	16.875	3.164	4.219	4.102
12 x 3	11-1/4 x 2-1/2	28.125	14.648	11.719	6.836
12 x 4	11-1/4 x 3-1/2	39.375	40.195	22.969	9.570
12 x 6	11-1/2 x 5-1/2	63.250	159.443	57.979	15.373
12 x 8	11-1/2 x 7-1/2	86.250	404.297	107.813	20.964
12 x 10	11-1/2 x 9-1/2	109.250	821.651	172.979	26.554
12 x 12	11-1/2 x 11-1/2	132.250	1457.505	253.479	32.144
12 x 14	11-1/2 x 13-1/2	155.250	2357.859	349.313	37.734
12 x 16	11-1/2 x 15-1/2	178.250	3568.713	460.479	43.325
12 x 18	11-1/2 x 17-1/2	201.250	5136.066	586.979	48.915
12 x 20	11-1/2 x 19-1/2	224.250	7105.922	728.813	54.505
12 x 22	11-1/2 x 21-1/2	247.250	9524.273	885.979	60.095
12 x 24	11-1/2 x 23-1/2	270.250	12437.129	1058.479	65.686
14 x 2	13-1/4 x 1-1/2	19.875	3.727	4.969	4.831
14 x 3	13-1/4 x 2-1/2	33.125	17.253	13.802	8.051
14 x 4	13-1/4 x 3-1/2	46.375	47.34	27.052	11.266
14 x 6	13-1/2 x 5-1/2	74.250	187.172	68.063	18.047
14 x 8	13-1/2 x 7-1/2	101.250	474.609	126.563	24.609
14 x 10	13-1/2 x 9-1/2	128.250	964.547	203.063	31.172
14 x 12	13-1/2 x 11-1/2	155.250	1710.984	297.563	37.734
14 x 14	13-1/2 x 13-1/2	182.250	2767.922	410.063	44.297
14 x 16	13-1/2 x 15-1/2	209.250	4189.359	540.563	50.859
14 x 18	13-1/2 x 17-1/2	236.250	6029.297	689.063	57.422
14 x 20	13-1/2 x 19-1/2	263.250	8341.734	855.563	63.984
14 x 22	13-1/2 x 21-1/2	290.250	11180.672	1040.063	70.547
14 x 24	13-1/2 x 23-1/2	317.250	14600.109	1242.563	77.109

TABLE 4-1. *(Continued)*

Nominal size b(inches)d	Standard dressed size (S4S) b(inches)d	Area of Section A	Moment of inertia I	Section modulus S	Approx. weight*
16 x 3	15-1/4 x 2-1/2	38.125	19.857	15.885	9.267
16 x 4	15-1/4 x 3-1/2	53.375	54.487	31.135	12.975
16 x 6	15-1/2 x 5-1/2	85.250	214.901	78.146	20.720
16 x 8	15-1/2 x 7-1/2	116.250	544.922	145.313	28.255
16 x 10	15-1/2 x 9-1/2	147.250	1107.443	233.146	35.790
16 x 12	15-1/2 x 11-1/2	178.250	1964.463	341.646	43.325
16 x 14	15-1/2 x 13-1/2	209.250	3177.984	470.813	50.859
16 x 16	15-1/2 x 15-1/2	240.250	4810.004	620.646	58.394
16 x 18	15-1/2 x 17-1/2	271.250	6922.523	791.146	65.929
16 x 20	15-1/2 x 19-1/2	302.250	9577.547	982.313	73.464
16 x 22	15-1/2 x 21-1/2	333.250	12837.066	1194.146	80.998
16 x 24	15-1/2 x 23-1/2	364.250	16763.086	1426.646	88.533
18 x 6	17-1/2 x 5-1/2	96.250	242.630	88.229	23.394
18 x 8	17-1/2 x 7-1/2	131.250	615.234	164.063	31.901
18 x 10	17-1/2 x 9-1/2	166.250	1250.338	263.229	40.408
18 x 12	17-1/2 x 11-1/2	201.250	2217.943	385.729	48.915
18 x 14	17-1/2 x 13-1/2	236.250	3588.047	531.563	57.422
18 x 16	17-1/2 x 15-1/2	271.250	5430.648	700.729	65.929
18 x 18	17-1/2 x 17-1/2	306.250	7815.754	893.229	74.436
18 x 20	17-1/2 x 19-1/2	341.250	10813.359	1109.063	82.943
18 x 22	17-1/2 x 21-1/2	376.250	14493.461	1348.229	91.450
18 x 24	17-1/2 x 23-1/2	411.250	18926.066	1610.729	99.957
20 x 6	19-1/2 x 5-1/2	107.250	270.359	98.313	26.068
20 x 8	19-1/2 x 7-1/2	146.250	685.547	182.813	35.547
20 x 10	19-1/2 x 9-1/2	185.250	1393.234	293.313	45.026
20 x 12	19-1/2 x 11-1/2	224.250	2471.422	429.813	54.505
20 x 14	19-1/2 x 13-1/2	263.250	3998.109	592.313	63.984
20 x 16	19-1/2 x 15-1/2	302.250	6051.297	780.813	73.464
20 x 18	19-1/2 x 17-1/2	341.250	8708.984	995.313	82.943
20 x 20	19-1/2 x 19-1/2	380.250	12049.172	1235.813	92.422
20 x 22	19-1/2 x 21-1/2	419.250	16149.859	1502.313	101.901
20 x 24	19-1/2 x 23-1/2	458.250	21089.047	1794.813	111.380
22 x 6	21-1/2 x 5-1/2	118.250	298.088	108.396	28.741
22 x 8	21-1/2 x 7-1/2	161.250	755.859	201.563	39.193
22 x 10	21-1/2 x 9-1/2	204.250	1536.130	323.396	49.644
22 x 12	21-1/2 x 11-1/2	247.250	2724.901	473.896	60.095
22 x 14	21-1/2 x 13-1/2	290.250	4408.172	653.063	70.547
22 x 16	21-1/2 x 15-1/2	333.250	6671.941	860.896	80.998
22 x 18	21-1/2 x 17-1/2	376.250	9602.211	1097.396	91.450
22 x 20	21-1/2 x 19-1/2	419.250	13284.984	1362.563	101.901
22 x 22	21-1/2 x 21-1/2	462.250	17806.254	1656.396	112.352
22 x 24	21-1/2 x 23-1/2	505.250	23252.023	1978.896	122.804
24 x 6	23-1/2 x 5-1/2	129.250	325.818	118.479	31.415
24 x 8	23-1/2 x 7-1/2	176.250	826.172	220.313	42.839
24 x 10	23-1/2 x 9-1/2	223.250	1679.026	353.479	54.262
24 x 12	23-1/2 x 11-1/2	270.250	2978.380	517.979	65.686
24 x 14	23-1/2 x 13-1/2	317.250	4818.234	713.813	77.109
24 x 16	23.1/2 x 15-1/2	364.250	7292.586	940.979	88.533
24 x 18	23-1/2 x 17-1/2	411.250	10495.441	1199.479	99.957
24 x 20	23-1/2 x 19-1/2	458.250	14520.797	1489.313	111.380
24 x 22	23-1/2 x 21-1/2	505.250	19462.648	1810.479	122.804
24 x 24	23-1/2 x 23-1/2	552.250	25415.004	2162.979	134.227

4-4 Transferring Moments of Inertia

In the design of glued built-up beams of lumber and plywood it becomes necessary to determine the moment of inertia of the total cross section. Figure 4-3*a* shows one type of cross section for such built-up members. To accomplish this we must transfer moments of inertia from one axis to another by making use of the *transfer-of-axis equation,* sometimes called the *transfer formula.* It may be stated thus: *The moment of inertia of a section about any axis parallel to an axis through its own centroid is equal to the moment of inertia of the section about its own gravity (centroidal) axis, plus its area times the square of the perpendicular distance between the two axes.* Expressed mathematically,

$$I = I_0 + Az^2$$

In this formula

I = moment of inertia of the section about the required axis,
I_0 = moment of inertia of the section about its own gravity (centroidal) axis parallel to the required axis,
A = area of the section,
z = distance between the two parallel axes.

Example. A built-up beam of the type shown in Fig. 4-3*a* has an overall depth of 32 in. [812 mm] and flange pieces consisting of two 8 × 6 [190 × 140 mm] timbers. Using the transfer formula, compute the moment of inertia of the flange timbers about the centroidal axis X–X.

Solution: (1) Figure 4-3*b* is constructed from the given data. Because of symmetry, the centroidal axis will occur at middepth 16 in. [406 mm] from the top.

(2) The gravity axis of each 8 × 6 piece occurs at its center, making $z = 16 - 2.75 = 13.25$ in. [336 mm].

(3) From Table 4-1, I_0 of one 8 × 6 piece is 104 in.^{4} [43.45×10^6 mm^4] and its area is $A = 41.25$ in.^{2} [26.6×10^3 mm^2].

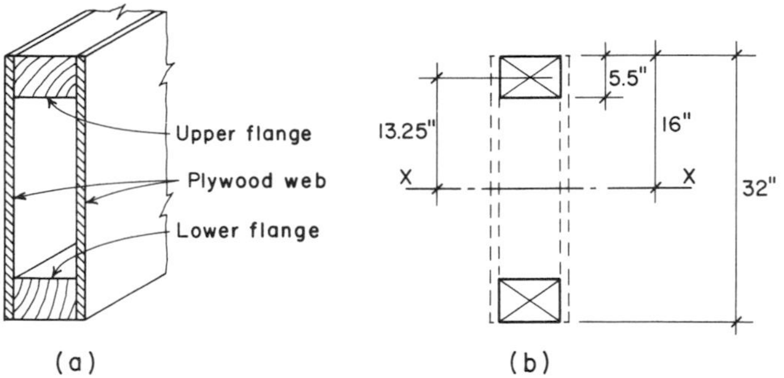

FIGURE 4-3.

(4) Substituting in the transfer formula, the moment of inertia of one of the 8 × 6 pieces is

$$I_X = I_0 + Az^2 = 104 + (41.25 \times 13.25^2)$$
$$= 7346 \text{ in.}^4 \ [3 \times 10^9 \text{ mm}^4]$$

(5) The moment of inertia of the lower flange piece is also 7346 in.^{4}, making the total I_X of both pieces equal to 2 × 7346 = 14,692 in.^{4}.

4-5 Section Modulus

One of the properties of sections used by the structural designer is called the *section modulus*. Its use in the design of beams is explained later (Chapters 7 and 9); for the present it is necessary to know only that if I is the moment of inertia of a section about an axis passing through the centroid and c is *the distance from the most remote edge of the section to the same axis,* the section modulus is equal to I/c. The letter S is used to denote section modulus. Because I is in units of inches to the fourth power (in.^{4}) and c is a linear dimension in inches, the section modulus $S = I/c$ is in units of inches to the third power (in.^{3}).

For the rectangular beam cross section shown in Fig. 4-2*a*, b designates the width of the section and d the depth. The distance from the most remote edge to axis $X–X$ is $c = d/2$. We know that I_{X-X} for the section is $bd^3/12$. Therefore the section modulus is

$$S = \frac{I}{c} = \frac{bd^3}{12} \div \frac{d}{2} = \frac{bd^3}{12} \times \frac{2}{d} \quad \text{or} \quad S = \frac{bd^2}{6}$$

It is rarely necessary to solve this formula because extensive tables that give the section modulus for various structural shapes are available. (See Table 4-1.)

Example. Find the section modulus of an 8 × 10 beam about an axis through the centroid parallel to the shorter side.

Solution: Referring to Table 4-1, we find that the dressed dimensions of this member are 7.5 by 9.5 in. The section modulus is given in the fourth column as 112.8 in.^{3} Verifying this value,

$$S = \frac{bd^2}{6} = \frac{7.5 \times 9.5 \times 9.5}{6} = 112.8 \text{ in.}^3$$

4-6 Radius of Gyration

This property of the cross section of a structural member is related to the design of compression members. It is dependent on the size and shape of the section and is an index of the stiffness of the section when used as a column or strut. The *radius of gyration* is defined mathematically as $r = \sqrt{I/A}$, where I is the moment of inertia and A the area of the section. It is expressed in inches because moment of inertia is in inches^4 and the cross-sectional area is in square inches. Radius of gyration is not used as widely in structural wood design as it is in the design of structural steel. The rectangular sections commonly employed for wood columns make it convenient to substitute the *least lateral dimension* for radius of gyration in column design procedures.

(*Note:* Use standard dressed sizes in the solution of the following problems unless otherwise stated.)

Problem 4-6-A. Verify by computation the value listed in Table 4-1 for the moment of inertia of a 4 × 12 plank with respect to a horizontal axis through the centroid parallel to the longer side.

Problem 4-6-B. If the plank in Problem 4-6-A is turned on edge, what is the moment of inertia with respect to the centroidal axis parallel to the shorter side?

Problem 4-6-C. Compute the moment of inertia of a pole with an actual diameter of 8 in., with respect to the centroidal axis of its circular cross section. Is this greater or less than the $I_{X\text{–}X}$ of a nominal 8 × 8 post?

Problem 4-6-D*. If the flange timbers in a lumber and plywood box beam (Fig. 4-3*a*) consist of 6 × 4 pieces with the 6 in. sides horizontal and the beam has an overall depth of 24 in., compute the moment of inertia of the flange pieces about the centroidal *X–X* axis of the build-up section.

Problem 4-6-E. Verify by computation the value listed in Table 4-1 for the section modulus of an 8 × 10 member about an axis through the centroid parallel to the longer side.

Problem 4-6-F. Find the radius of gyration of the pole described in Problem 4-6-C.

5

Beam Functions

This chapter deals with the external and internal force effects in beams. For the principles being illustrated, the actual units of loads and dimensions are less important than their relationships. For this reason, in order to reduce confusion in the computations, we have omitted the use of dual units and present work only in U.S. units. For those who wish to work the exercise problems in SI units, however, we have given both sets of units.

5-1 Introduction

A beam is a structural member that resists transverse loads. Generally, the loads act at right angles to the longitudinal axis of the beam. The loads on a beam tend to *bend* it, and we say the member is in *flexure* or *bending*. The supports for beams are usually at or near the ends, and the supporting upward forces are called *reactions*. A beam that supports smaller beams is called a *girder*.

A *simple beam* rests on a support at each end, the beam ends being free to rotate. The majority of beams in wood construction are simple beams (Fig. 5-1*a*).

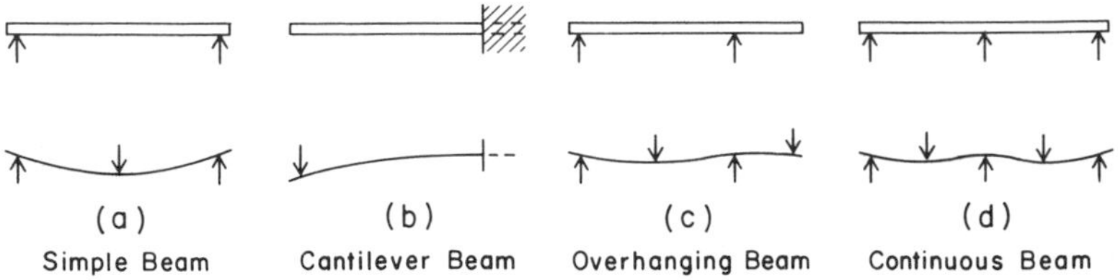

FIGURE 5-1.

A *cantilever beam* projects beyond its support. A beam embedded in a wall and extending beyond the face of the wall is a typical example (Fig. 5-1*b*). This beam is said to be *fixed* or *restrained* at the support.

An *overhanging beam* projects beyond one or both of its supports (Fig. 5-1*c*).

A *continuous beam* is a beam that rests on three or more supports (Fig. 5-1*d*).

In this chapter we are concerned with the external forces acting on beams and the measurement of their effect in terms of shear and bending moment.

5-2 Moment of a Force

The *moment* of a force is its tendency to cause rotation about a given point or axis. The magnitude of the moment is equal to the magnitude of the force multiplied by the perpendicular distance from its line of action to the point about which the moment is taken. Moments are expressed in compound units such as foot-pounds and inch-pounds or kip-feet and kip-inches. The point or axis about which the force tends to cause turning is called the *center of moments*, and the perpendicular distance between the line of action of the force and the center of moments is called the *lever arm* or *moment arm*.

Consider the cantilever beam 10 ft in length shown in Fig. 5-2. A force of 500 lb is placed 4 ft 0 in. from the face of the wall. The moment of this force *about point A* at the face of the wall is $500 \times 4 = 2000$ ft-lb. If the force is moved to the free (unsupported) end

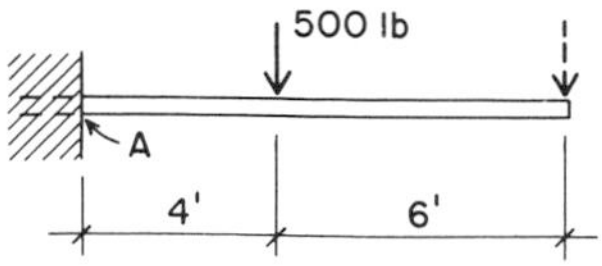

FIGURE 5-2.

of the beam, as indicated by the dotted line, the moment of the force about point A becomes $500 \times 10 = 5000$ ft-lb.

5-3 Determination of Reactions

The reactions of a beam are determined by applying the three laws of static equilibrium, which may be expressed as follows:

1. The algebraic sum of all vertical forces equals zero.
2. The algebraic sum of all horizontal forces equals zero.
3. The algebraic sum of the moments of all forces about any point equals zero.

In a beam that supports vertical loads only there will be no horizontal forces involved. The reactions will therefore be vertical and, in the case of symmetrically loaded, simple beams, each reaction will be equal to half the sum of the loads. It is convenient to distinguish between the two reactions by using the symbols R_1 and R_2 to represent the left and right reactions, respectively.

Consider first a simple beam with a span of 20 ft 0 in. and a concentrated load of 2400 lb at the center of the span, as indicated in Fig. 5-3a. Because this beam is symmetrically loaded and the

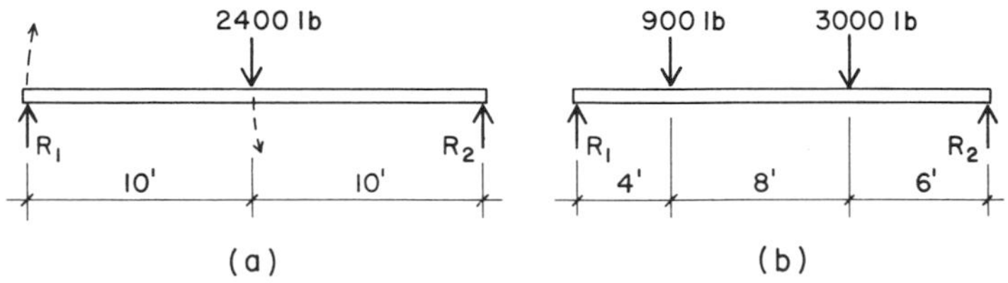

FIGURE 5-3.

downward force is 2400 lb, it is obvious that each reaction is equal to one-half the load, or 1200 lb. R_1 and R_2 each equals 1200 lb. We can prove this by employing the third law of equilibrium as previously stated. Let us take a point at R_2, the right reaction, as the center of moments. The force R_1 tends to produce a clockwise rotation about this point, and its moment is $R_1 \times 20$. The downward force, the load, tends to cause a counterclockwise rotation about the same point, and its moment is 2400×10. Note the arrows that show the directions of rotation. Because the sum of the moments that tend to cause clockwise rotation equals the sum of the moments that tend to cause counterclockwise rotation, we may write

$$R_1 \times 20 = 2400 \times 10$$
$$20R_1 = 24{,}000$$
$$R_1 = 1200 \text{ lb}$$

In a similar manner we can show that $R_2 = 1200$ lb.

Now let us consider the simple beam with two loads, as indicated in Fig. 5-3*b*. This beam is not symmetrically loaded; hence the reactions are not necessarily equal. Let us compute their magnitudes.

We select a point on the line of action of R_2 as the center of moments. With respect to this point, the only force that tends to cause a clockwise rotation is R_1, and its moment is $R_1 \times 18$. For the same center of moments the loads 900 and 3000 lb tend to produce counterclockwise rotation, and their moments are (900×14) and (3000×6), respectively. Then, in accordance with the law,

$$18 \times R_1 = (900 \times 14) + (3000 \times 6)$$
$$18R_1 = 30{,}600$$
$$R_1 = 1700 \text{ lb}$$

Taking R_1 as a center of moments, we may write

$$18 \times R_2 = (900 \times 4) + (3000 \times 12)$$
$$18R_2 = 39{,}600$$
$$R_2 = 2200 \text{ lb}$$

The magnitudes of the reactions just computed may be readily checked by employing the first law of equilibrium. In effect, this law states that the sum of the downward forces equals the sum of the upward forces or *the sum of the loads equals the sum of the reactions*. Thus

$$900 + 3000 = 1700 + 2200$$

or

$$3900 \text{ lb} = 3900 \text{ lb}$$

The discussion so far concerns the computation of the reactions for a simple beam with concentrated loads. Now let us consider a beam with a uniformly distributed load and a concentrated load. *As far as the reactions are concerned,* a distributed load has the same effect as a concentrated load of the same magnitude acting at the center of gravity of the distributed load.

Example 1. Compute the reactions for the beam and loading shown in Fig. 5-4*a*.

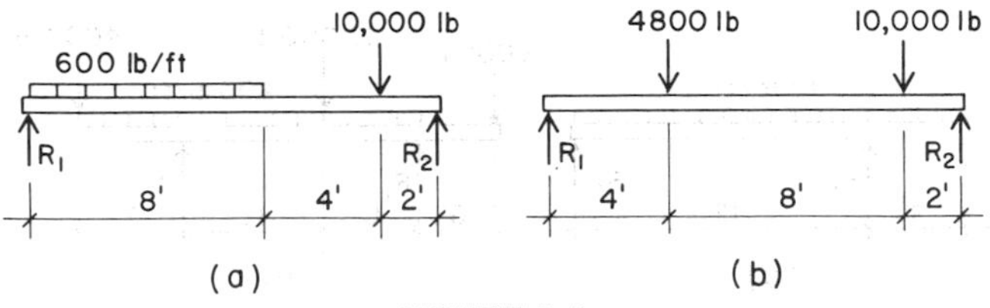

FIGURE 5-4.

Solution: This beam has a concentrated load of 10,000 lb and a uniformly distributed load of 600 lb per lin ft extending over a length of 8 ft 0 in. The magnitude of the distributed load is 8×600 or 4800 lb; its center of gravity lies at 4 ft 0 in. from R_1 and at 10 ft 0 in. from R_2. With respect to the reactions, the beams and loads shown in Figs. 5-4*a* and *b* are similar.

Consider Fig. 5-4*a* and take R_2 as the center of moments. Then

$$14 \times R_1 = (8 \times 600 \times 10) + (10{,}000 \times 2)$$

$$R_1 = 4857.1 \text{ lb}$$

In this equation 8×600 is the magnitude of the distributed load and 10 is its lever arm about R_2.

Taking R_1 as the center of moments,

$$14 \times R_2 = (8 \times 600 \times 4) + (10{,}000 \times 12)$$

$$R_2 = 9942.9 \text{ lb}$$

To check,

$$(8 \times 600) + 10{,}000 = 4857.1 + 9942.9$$

$$14{,}800 \text{ lb} = 14{,}800 \text{ lb}$$

Example 2. Compute the reactions for the overhanging beam with a uniformly distributed load, as shown in Fig. 5-5*a*.

Solution: By observation we note that the center of gravity of the uniformly distributed load lies at a point 8 ft 0 in. from R_2 and

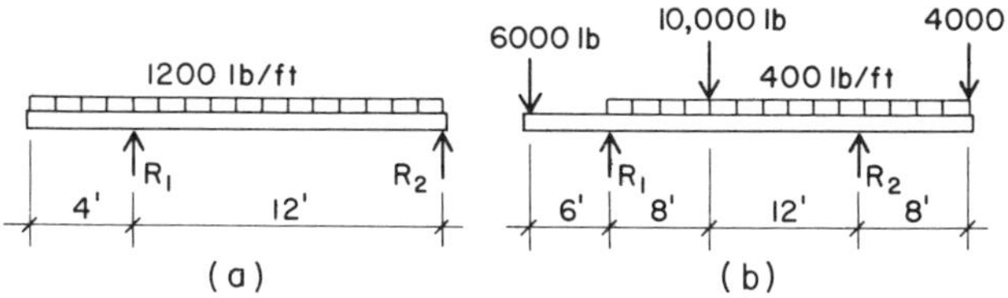

FIGURE 5-5.

4 ft 0 in. to the right of R_1. Then, selecting R_2 as the center of moments,

$$12 \times R_1 = (16 \times 1200 \times 8)$$
$$R_1 = 12{,}800 \text{ lb}$$

With R_1 as the center of moments,

$$12 \times R_2 = (16 \times 1200 \times 4)$$
$$R_2 = 6400 \text{ lb}$$

Check:

$$(16 \times 1200) = 12{,}800 + 6400$$
$$19{,}200 \text{ lb} = 19{,}200 \text{ lb}$$

Example 3. Compute the reactions for the overhanging beam and loading shown in Fig. 5-5*b*.

Solution: Study of the figure will show that the distributed load has its center of gravity 6 ft to the left of R_2 and 14 ft to the right of R_1. Selecting R_2 as the center of moments,

$$(20 \times R_1) + (4000 \times 8) = (6000 \times 26) + (10{,}000 \times 12) + (28 \times 400 \times 6)$$
$$20R_1 = 311{,}200$$
$$R_1 = 15{,}560 \text{ lb}$$

Taking R_1 as the center of moments,

$$(20 \times R_2) + (6000 \times 6) = (10{,}000 \times 8) + (4000 \times 28) + (28 \times 400 \times 14)$$
$$20R_2 = 312{,}800$$
$$R_2 = 15{,}640 \text{ lb}$$

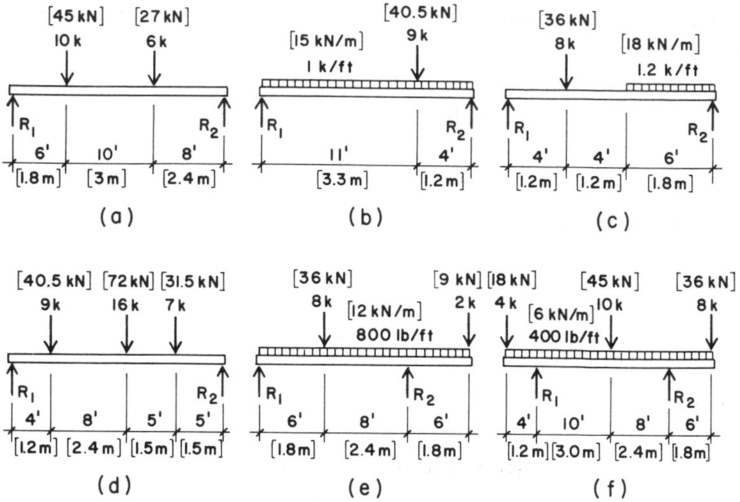

FIGURE 5-6.

Check:

$$6000 + 10{,}000 + 4000 + (28 \times 400) = 15{,}560 + 15{,}640$$

$$31{,}200 \text{ lb} = 31{,}200 \text{ lb}$$

Problems 5-3-A-B-C-D-E*-F. Compute the magnitudes of the reactions for the beams and loadings shown in Fig. 5-6. Use two moment equations and check your results by summation of the vertical forces.

5-4 Vertical Shear

The tendency for a beam to fail by dropping vertically between supports was mentioned in Sec. 2-4 and illustrated in Fig. 2-3*b*. This tendency for one part of a beam to move vertically with respect to an adjacent part is called *vertical shear*. The magnitude of the vertical shear at any section along the length of a beam is the algebraic sum of the vertical forces on either side of the section. If we call the upward forces (reactions) positive and the downward forces (loads) negative, we may say that *the vertical shear at any section of a beam is equal to the reactions minus the* *loads to the left of the section*. If we take the forces to the right of the section instead of the left, the magnitude of the shear will be the same. To avoid confusion, however, we shall consider the forces to the left in the illustrative examples. The magnitude of the vertical shear is represented by V. If the loads and reactions are in units of pounds or kips, the vertical shear will also be in units of pounds or kips.

There are two important reasons for investigating the vertical shear in beams: (1) it is necessary to know the maximum value of the shear and (2) it is necessary to locate the section at which the shear changes from a positive to a negative quantity—the section at which the shear "passes through zero." At this section the tendency of a beam to fail in bending is greatest.

A *shear diagram* is a graphical representation of the values of the vertical shear throughout the length of a beam. A horizontal line (the base line) is drawn directly below the beam diagram, and values of the shear at various sections along the span are plotted to a convenient scale: positive values above and negative values below the base line.

Example 1. A simple beam 20 ft long carries two concentrated loads, as indicated in Fig. 5-7*a*. Construct the shear diagram.

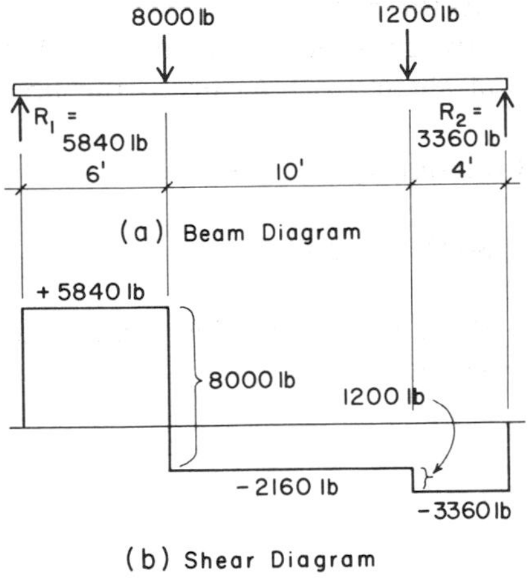

FIGURE 5-7.

Solution: (1) Compute the reactions as explained in Sec. 5-3.

$$20R_1 = (8000 \times 14) + (1200 \times 4)$$

$$20R_1 = 116{,}800$$

$$R_1 = 5840 \text{ lb}$$

$$20R_2 = (8000 \times 6) + (1200 \times 16)$$

$$20R_2 = 67{,}200$$

$$R_2 = 3360 \text{ lb}$$

(2) To designate the section at which we wish to compute the value of the shear, it is convenient to use a subscript: $V_{(x=4)}$. This indicates that the value of the shear is taken at a section 4 ft from the left end of the beam.

First consider a section 1 ft to the right of R_1. Because the shear is equal to the reactions minus the loads to the left of the section and the reaction to the left is 5840 lb and there are no loads to the left, we may write

$$V_{(x=1)} = 5840 - 0 = 5840 \text{ lb}$$

This is a positive quantity, and a point is plotted at a convenient scale above the base line at 1 ft to the right of R_1.

Note that there are no loads between R_1 and the load of 8000 lb 6 ft from R_1. Hence the vertical shear from R_1 up to the first concentrated load is 5840 lb.

Next consider the section 8 ft from R_1. We write

$$V_{(x=8)} = 5840 - 8000 = -2160 \text{ lb}$$

This is a negative value and is plotted *below* the base line. The value of the shear does not change between the two concentrated loads. In the same manner,

$$V_{(x=18)} = 5840 - (8000 + 1200) = -3360 \text{ lb}$$

This is the value of the shear at all sections between the 1200-lb load and R_2. Thus the shear diagram is completed. (See Fig. 5-7*b*.) Note that all the vertical distances (ordinates) in the shear diagram show the values of the vertical shear for all sections of the beam.

(3) Having completed the shear diagram, let us see what it discloses. First we note that the maximum value of the vertical shear is 5840 lb. It occurs at all sections between the left reaction and the concentrated load of 8000 lb. It is obvious that for simple beams the maximum vertical shear occurs at the greater reaction and is equal to the greater reaction in magnitude. We see also that the value of the shear changes sign (from positive values to negative values) directly under the 8000 lb load, at 6 ft from R_1. Later it is shown that this is the section at which the maximum bending moment occurs. This value is critical in the design of beams.

It should be noted that the weight of a beam constitutes a uniformly distributed load. Because the beam weight is often quite small when compared with the loads to be carried, it is sometimes neglected in the computations. In this chapter consideration of beam weight has been omitted in order to focus on the separate effects of concentrated and uniform loads.

Example 2. The simple beam shown in Fig. 5-8*a* has both a concentrated and a uniformly distributed load. Construct the

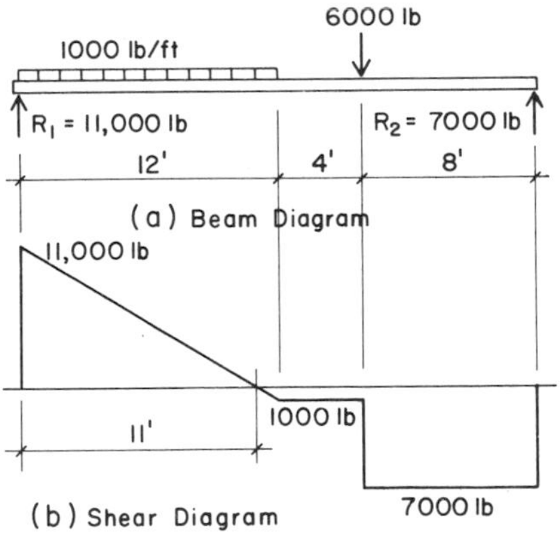

FIGURE 5-8.

shear diagram, designate the value of the maximum shear, and locate the section at which the shear passes through zero.

Solution: (1) Computing the reactions,

$$24R_1 = (12 \times 1000 \times 18) + (6000 \times 8)$$
$$24R_1 = 264{,}000$$
$$R_1 = 11{,}000 \text{ lb}$$
$$24R_2 = (12 \times 1000 \times 6) + (6000 \times 16)$$
$$24R_2 = 168{,}000$$
$$R_2 = 7000 \text{ lb}$$

(2) The value of the shear at the left reaction is 11,000 lb. Noting that the magnitude of the uniformly distributed load is 1000 lb per lin ft, we may compute the shear value at essential points:

$$V_{(x=1)} = 11{,}000 - (1 \times 1000) = 10{,}000 \text{ lb}$$
$$V_{(x=2)} = 11{,}000 - (2 \times 1000) = 9000 \text{ lb}$$
$$V_{(x=12)} = 11{,}000 - (12 \times 1000) = -1000 \text{ lb}$$
$$V_{(x=16-)} = 11{,}000 - (12 \times 1000) = -1000 \text{ lb}$$
$$V_{(x=16+)} = 11{,}000 - [(12 \times 1000) + 6000] = -7000 \text{ lb}$$
$$V_{(x=24)} = 11{,}000 - [(12 \times 1000) + 6000] = -7000 \text{ lb}$$

In these equations $V_{(x=16-)}$ indicates a section close to but not including the 6000-lb load. Also, $V_{(x=16+)}$ designates a section slightly to the right of the 6000-lb load. Note that under the uniformly distributed load the shear diagram is a sloping line. In plotting the shear for a beam that has only a uniformly distributed load, it is necessary to compute only the shear values at the two ends of the distributed load.

(3) From the shear diagram in Fig. 5-8*b* it is observed that the value of the maximum vertical shear is 11,000 lb; it occurs at the left reaction. We notice also that the shear passes through zero at some point between the left end of the beam and the end of the

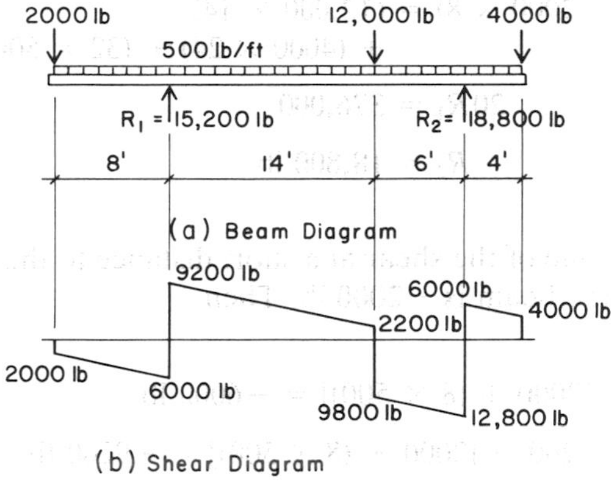

FIGURE 5-9.

distributed load. To find the exact location of this section let us called it x feet from R_1. Next, we write an expression for the shear at this section and equate it to zero because the value of V at this section is zero. Then

$$11{,}000 - (1000 \times x) = 0$$
$$1000x = 11{,}000$$

and $x = 11$ ft, the section at which the shear passes through zero.

Example 3. Figure 5-9*a* shows a beam overhanging both ends. A uniformly distributed load of 500 lb per lin ft extends over the entire length of the beam. In addition, there are three concentrated loads located at the positions shown. Construct the shear diagram, designate the value of the maximum shear, and locate the section at which the shear passes through zero.

Solution: (1) Computing the reactions,

$$20R_1 + (4000 \times 4) = (12{,}000 \times 6) + (2000 \times 28) + (32 \times 500 \times 12)$$
$$20R_1 = 304{,}000$$
$$R_1 = 15{,}200 \text{ lb}$$

$$20R_2 + (2000 \times 8) = (12{,}000 \times 14) + (4000 \times 24) + (32 \times 500 \times 8)$$

$$20R_1 = 376{,}000$$

$$R_2 = 18{,}800 \text{ lb}$$

(2) The value of the shear at a short distance to the right of the left end of the beam is −2000 lb. Then

$$V_{(x=8-)} = -[2000 + (8 \times 500)] = -6000 \text{ lb}$$

$$V_{(x=8+)} = 15{,}200 - [2000 + (8 \times 500)] = +9200 \text{ lb}$$

$$V_{(x=22-)} = 15{,}200 - [2000 + (22 \times 500)] = +2200 \text{ lb}$$

$$V_{(x=22+)} = 15{,}200 - [2000 + 12{,}000 + (22 \times 500)] = -9800 \text{ lb}$$

$$V_{(x=28-)} = 15{,}200 - [2000 + 12{,}000 + (28 \times 500)] = -12{,}800 \text{ lb}$$

$$V_{(x=28+)} = (15{,}200 + 18{,}800) - [2000 + 12{,}000 + (28 \times 500)] = +6000 \text{ lb}$$

$$V_{(x=32-)} = (15{,}200 + 18{,}800) - [2000 + 12{,}000 + (32 \times 500)] = +4000 \text{ lb}$$

(3) The shear diagram is constructed in accordance with these values and is shown in Fig. 5-9*b*. In reading the maximum value of the shear, we disregard the positive or negative signs, for the diagrams are merely conventional methods of representing the absolute numerical values. Therefore the maximum vertical shear is 12,800 lb and occurs immediately to the left of the right reaction. Note also that the shear passes through zero at three different points along the span: at R_1, at R_2, and under the 12,000-lb load. The significance of these multiple crossings of the base line is discussed in Sec. 5-5.

Problems 5-4-A-B-C-D-E*-F. For the beams and loadings shown in Fig. 5-10 draw the shear diagrams, designate the values of the maximum vertical shear, and locate the sections at which the shear passes through zero.

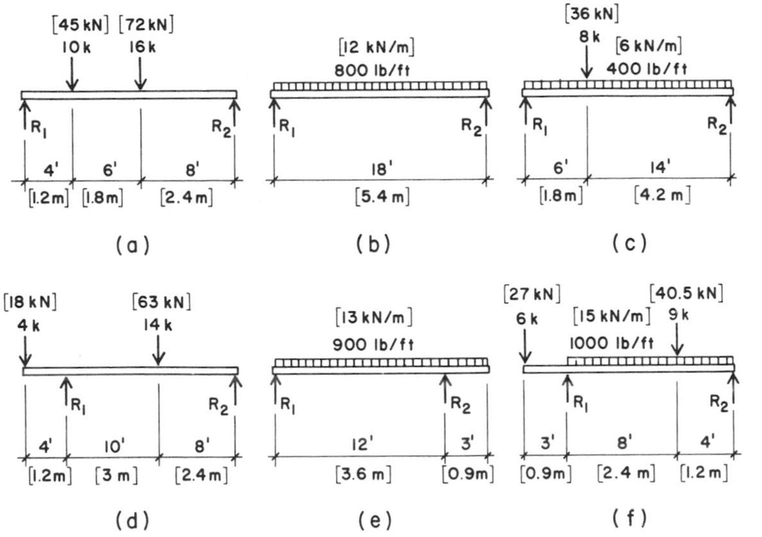

FIGURE 5-10.

5-5 Bending Moment

The *bending moment* at any section in the length of a beam is the measure of the tendency of the beam to bend due to the forces acting on it. The magnitude of the bending moment varies throughout the length of a beam, with the maximum value occurring at the section at which the shear passes through zero. Bending moment is represented by the letter M.

The magnitude of the bending moment at any section of a beam is equal to the algebraic sum of the moments of the forces on either the right or left of the section. For the purpose of simplification let us consider only the forces to the *left* of the section. We may then say *the bending moment at any section in the length of a beam equals the moments of the reactions minus the moments of the loads to the left of the section.*

Particular attention is called to the similarity between this rule and the rule for determining the vertical shear given in Sec. 5-4. Shear and bending moment are often confused. Remember that the shear is *reactions minus loads,* whereas the bending moment is *moments of reactions minus moments of loads.* Because a moment is the result of multiplying a force by a distance, bending moments are expressed in foot-pounds and inch-pounds (or kip-feet and kip-inches).

To illustrate let us consider the simple beam shown in Fig. 5-11*a* which supports a concentrated load P at the center of its span L. Because the beam is symmetrically loaded, each reaction is $P/2$ and the shear diagram may be constructed as shown. The value of the maximum shear is $P/2$ and the shear passes through zero at the center of the span. At this section the bending moment will be maximum. Applying the rule stated, we note that the reaction to the left is $P/2$ and its lever arm is $L/2$; therefore the moment of the reaction is $(P/2) \times (L/2)$. There are no loads to the left of the section. If load P is considered, it has a lever arm of zero and its moment is $P \times 0 =$ zero. Therefore the maximum bending moment is

$$M_{(x=L/2)} = \frac{P}{2} \times \frac{L}{2} = \frac{PL}{4}$$

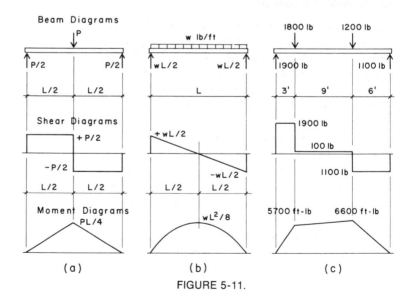

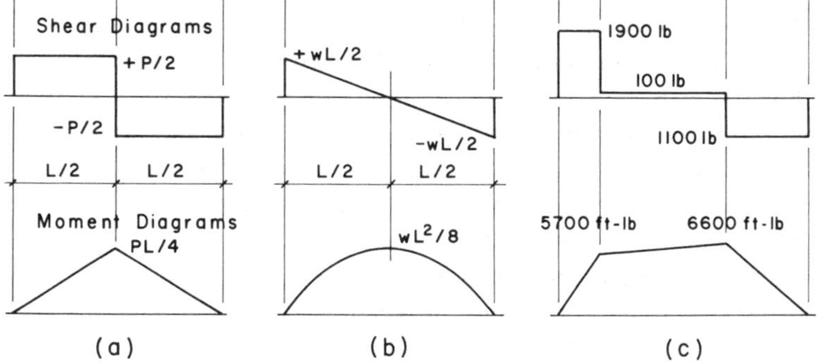

FIGURE 5-11.

The bending moment diagram is constructed as shown in the figure. Because a simple beam with a concentrated load at the center of the span is a condition that occurs frequently in practice, it will be well to remember this formula for maximum bending moment. (See Case I, Fig. 5-13.)

Another typical case is a simple beam with a uniformly distributed load that extends over the full length of the beam. Let L be the span length and w, the magnitude of the uniformly distributed load per linear foot of span, as indicated in Fig. 5-1*b*. The loading is symmetrical and the total distributed load is $w \times L$, which makes each reaction equal to $wL/2$. The maximum vertical shear is $wL/2$, and we note that the shear passes through zero at the center of the span. Computing the maximum value of the bending moment, which occurs at this section, we note that the reaction to the left is $wL/2$ and its lever arm is $L/2$. The load to the left of the section is $wL/2$ and its lever arm (because it acts at its center of gravity) is $L/4$. Then

$$M_{(x=L/2)} = \left(\frac{wL}{2} \times \frac{L}{2}\right) - \left(\frac{wL}{2} \times \frac{L}{4}\right)$$

$$M = \frac{wL^2}{4} - \frac{wL^2}{8}$$

$$M = \frac{wL^2}{8}$$

In the foregoing discussion the uniform load was expressed as w lb per lin ft. It is frequently convenient, however, to work with the *total* uniformly distributed load W, where $W = w \times L$. Accordingly, the maximum bending moment may be computed by either of the formulas

$$M = \frac{wL^2}{8} \quad \text{or} \quad M = \frac{WL}{8}$$

These values should be remembered also, for they will be used many times. The bending moment diagram for a simple beam with a uniformly distributed load takes the form of a parabola, as shown in the figure.

Example 1. A simple beam has a span of 18 ft with a concentrated load of 8 kips at the center of the span. Compute the maximum bending moment.

Solution: We have found that the maximum bending moment for this beam and loading occurs at the center of the span and is equal to $PL/4$. Therefore

$$M = \frac{PL}{4} = \frac{8 \times 18}{4} = 36 \text{ kip-ft}$$

In the design of beams it is generally necessary that the bending moment be expressed in kip-inches or inch-pounds. To convert kip-feet to kip-inches we merely multiply the magnitude in kip-feet by 12. Thus 36 kip-ft = 36 × 12 = 432 kip-in.

Example 2. A simple beam has a span of 22 ft and carries a uniformly distributed load of 300 lb per lin ft over its entire length. Compute the maximum bending moment.

Solution: For this typical case we have found that the maximum bending moment is $M = wL^2/8$. Therefore

$$M = \frac{wL^2}{8} = \frac{300 \times 22 \times 22}{8} = 18{,}150 \text{ ft-lb} \quad \text{or} \quad 217{,}800 \text{ in.-lb}$$

If the total uniformly distributed load had been given instead of the uniform load per foot, W would then have been 300 × 22 = 6600 lb and the bending moment computation would have become

$$M = \frac{WL}{8} = \frac{6600 \times 22}{8} = 18{,}150 \text{ ft-lb} \quad \text{or} \quad 217{,}800 \text{ in.-lb}$$

Example 3. Compute the maximum bending moment for the simple beam with two concentrated loads shown in Fig. 5-11*c*.

Solution: (1) To compute the maximum bending moment we must first determine its location; this necessitates constructing the shear diagram after the reactions are found.

(2) Computing the reactions,

$$18R_1 = (1800 \times 15) + (1200 \times 6)$$

$$18R_1 = 34{,}200$$

$$R_1 = 1900 \text{ lb}$$

$$18R_2 = (1800 \times 3) + (1200 \times 12)$$

$$18R_2 = 19{,}800$$

$$R_2 = 1100 \text{ lb}$$

(3) The shear diagram is constructed as explained in Sec. 5-4. It is shown in Fig. 5-11*c*, and we find that the shear passes through zero under the 1200-lb load at the section 12 ft 0 in. from the left end of the beam.

(4) Computing the maximum bending moment at the section at which the shear passes through zero,

$$M_{(x=12)} = (1900 \times 12) - (1800 \times 9) = 6600 \text{ ft-lb}$$

(5) To construct the bending moment diagram we need only to compute the bending moment under the 1800-lb load:

$$M_{(x\text{-}3)} = 1900 \times 3 = 5700 \text{ ft-lb}$$

The bending moment diagram is shown in Fig. 5-11*c*.

Negative Bending Moment. In all examples discussed so far we have dealt with simple beams subjected to positive bending moment only; the entire moment diagram has been plotted above the base line. Such beams tend to bend under the loading, and the curve assumed by the bent beam is *concave upward*. Under this condition the fibers in the upper part of the beam are in compression. When a beam projects beyond a support (Fig. 5-12*a*), this

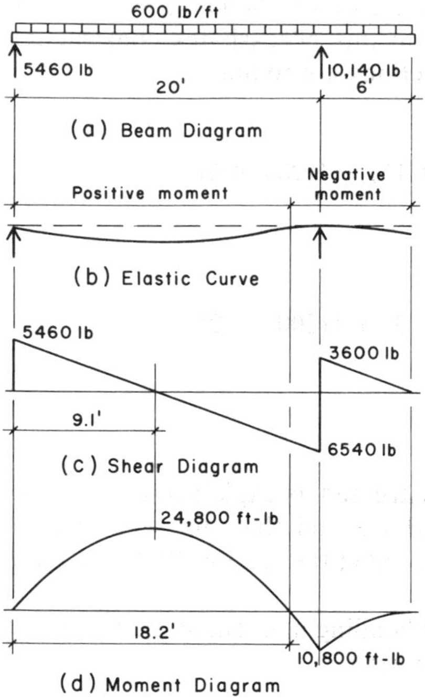

FIGURE 5-12.

segment of the beam has tensile stresses in its upper part. The bending moment for this condition is called negative, and the beam is bent *concave downward*. Figure 5-12*b* shows that there will be both positive and negative bending moment in an overhanging beam. The bending moment is zero at the section at which the curvature changes from concave upward to concave downward; this is called the *point of inflection*. If we construct moment diagrams, following the method previously described, the positive and negative bending moments will be shown graphically, and we may locate the point of inflection.

Example 4. Compute the maximum bending moment for the overhanging beam shown in Fig. 5-12*a* and draw the shear and bending moment diagrams. A uniformly distributed load of 600 lb per lin ft extends over the entire length of the beam.

Solution: (1) This is another situation where we must find the section at which the shear passes through zero in order to compute the magnitude of the maximum bending moment.

(2) Computing the reactions,

$$20R_1 = (26 \times 600 \times 7)$$
$$20R_1 = 109{,}200$$
$$R_1 = 5460 \text{ lb}$$
$$20R_2 = (26 \times 600 \times 13)$$
$$20R_2 = 202{,}800$$
$$R_2 = 10{,}140 \text{ lb}$$

(3) The shear diagram is constructed as shown in Fig. 5-12*c*. We find that the maximum vertical shear is 6540 lb and that the shear passes through zero at two places: one at some point between the two reactions and the other directly above the right reaction. To find the exact position of the first section that we will call x ft from R_1, write an expression for the shear at this section and equate it to zero:

$$5460 - (600 \times x) = 0$$
$$600x = 5460$$
$$x = 9.1 \text{ ft}$$

Then

$$M_{(x=9.1)} = (5460 \times 9.1) - (600 \times 9.1 \times 4.55) = +24{,}843 \text{ ft-lb}$$
$$M_{(x=20)} = (5460 \times 20) - (600 \times 20 \times 10) = -10{,}800 \text{ ft-lb}$$

It should be noted that the bending moment at 9.1 ft from R_1 is positive and the moment at R_2 is negative.

(4) To find the position of the inflection point let us call it x ft from R_1. For this section we write an expression for the bending moment and equate it to zero:

$$(5460 \times x) - \left(600 \times x \times \frac{x}{2}\right) = 0$$

$$\frac{600x^2}{2} - 5460x = 0$$

$$x^2 - 18.2x = 0$$

Completing the square,

$$x^2 - 18.2x + 82.81 = 82.81$$

Extracting the square root of both sides,

$$x - 9.1 = 9.1$$

$$x = 18.2 \text{ ft}$$

The position of the inflection point for steel and wood beams is relatively unimportant, but it is of vital importance that its position be known in the design of reinforced concrete.

(5) For this overhanging beam the maximum bending moment is 24,843 ft-lb; the negative bending moment is of lesser magnitude. In the design of such a beam the bending moment having the greater numerical value is the moment used in computations, regardless of whether it is positive or negative.

Problems 5-5-A-B-C-D-E*-F. For the beams and loadings shown in Fig. 5-10 draw the shear and bending moment diagrams. In each instance designate the maximum vertical shear and the maximum bending moment.

Problem 5-5-G*. For the beam shown in Fig. 5-10*f* determine the position of the inflection point.

5-6 Typical Loadings: Formulas for Beam Behavior

Some of the most common beam loadings are shown in Fig. 5-13. Shear and moment diagrams are shown for each loading and formulas are given for the values of reactions, maximum shear, maximum bending, and maximum deflection. (Deflection is discussed in Chapter 8.) For beam design, it is often sufficient to use only

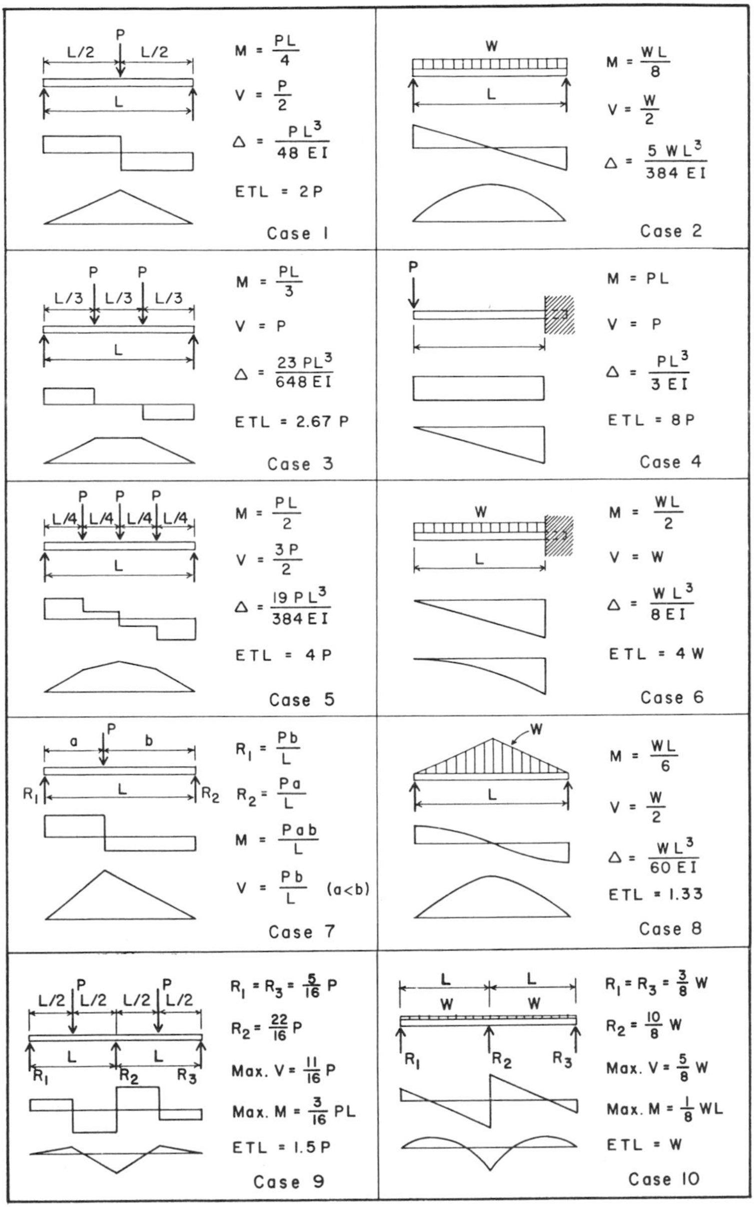

FIGURE 5-13. Values for typical beam loadings.

the values for maximum shear, bending, and deflection—especially for wood beams. Thus it is common for designers to use the formulas for these critical values for typical loadings, and to shorten the procedure for beam investigation.

More extensive loading conditions and accompanying formulas are given in various references, including Refs. 2 and 4 in the References following Chapter 20.

For a simple beam with a uniformly distributed load, the maximum moment is $WL/8$. For a simple beam with two equal loads at the third points of the span, the maximum moment is $PL/3$. These values are shown in Fig. 5-13, Cases 2 and 3, respectively. Equating these moment values, we can write

$$\frac{WL}{8} = \frac{PL}{3} \qquad \text{and} \qquad W = 2.67 \times P$$

which demonstrates that a total uniformly distributed load 2.67 times the value of one of the concentrated loads will produce the same maximum moment on the beam as that caused by the concentrated loads.

Coefficients for other loadings are also given in Fig. 5-13. It is important to remember that an ETL loading does not include the weight of the beam for which an estimated amount should be added. Also the shear values and reactions for the beam must be determined from the actual loading and not from the ETL.

5-7 Continuous Beams

As noted in Sec. 5-1, a continuous beam is a beam that rests on three or more supports. The magnitudes of the reactions cannot be determined solely by the principle of moments used for beams with only two supports. For this reason continuous beams are said to be "statically indeterminate," and a discussion of the necessary computations is beyond the scope of this book. Continuous beams are common in reinforced concrete structures and in welded steel construction but occur less frequently in wood construction.

Two continuous beam conditions that sometimes occur in wood construction are illustrated by Cases 9 and 10 in Fig. 5-13.

Case 10 shows a beam continuous over two equal spans with equal uniformly distributed loads W on each span. Case 9 has equal concentrated loads P at the centers of two equal continuous spans. In both cases formulas are given for finding the reactions, the maximum vertical shear and the maximum bending moment. For these two cases the maximum bending moment occurs over the central support and is negative. The maximum positive moments for Cases 10 and 9 (not shown in Fig. 5-13) are $M = WL/14.2$ and $5PL/32$, respectively. As noted earlier, in the design of beams the maximum numerical value is considered, regardless of whether it is positive or negative.

Example. A continuous beam has two equal spans of 10 ft each and a uniformly distributed load of 10,000 lb extending over each span. Construct the shear and bending moment diagrams and note the magnitudes of the maximum shear and maximum bending moment. The beam diagram is shown in Fig. 5-14*a*. Note that there will be two inflection points.

Solution: (1) To determine the reactions we refer to Case 10 in Fig. 5-13 and note that $R_1 = R_3 = \frac{3}{8}W$ and $R_2 = \frac{10}{8}W$. Then

$$R_1 = R_3 = \frac{3 \times W}{8} = \frac{3 \times 10{,}000}{8} = 3750 \text{ lb}$$

and

$$R_2 = \frac{10 \times W}{8} = \frac{10 \times 10{,}000}{8} = 12{,}500 \text{ lb}$$

(2) Values of the vertical shear are computed at various sections of the beam, as explained in Sec. 5-4, and the shear diagram is plotted as shown in Fig. 5-14*c*. For the conditions of this example Case 10 of Fig. 5-13 shows that the maximum vertical shear is $V = \frac{5}{8}W$. Thus

$$V = \frac{5 \times W}{8} = \frac{5 \times 10{,}000}{8} = 6250 \text{ lb}$$

This value occurs on both sides of the center support.

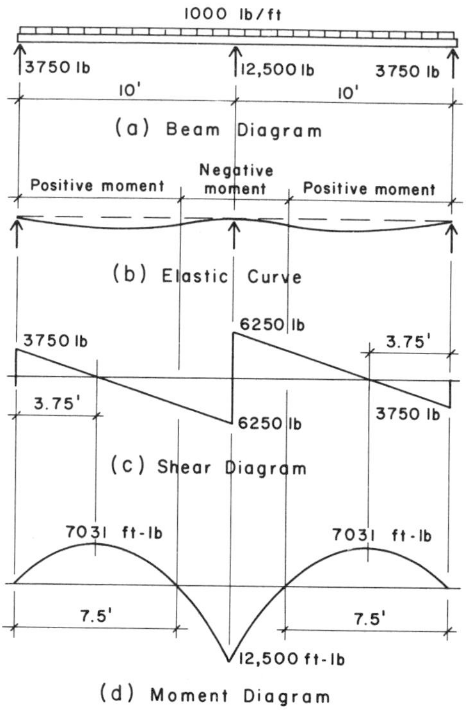

FIGURE 5-14.

(3) The shear passes through zero at the center reaction and at two other places. In the left span zero shear occurs at a distance $x = R_1/w = \frac{3750}{1000} = 3.75$ ft from R_1, w being equal to 1000 lb per lin ft. By symmetry the section of zero shear in the right span lies 3.75 ft from R_3.

(4) For a continuous beam of two equal spans with uniformly distributed loads the maximum bending moment occurs over the center support and is negative. Its value is $M = WL/8$, as shown in Fig. 5-13, Case 10. Then

$$M = \frac{WL}{8} = \frac{10{,}000 \times 10}{8} = 12{,}500 \text{ ft-lb}$$

This value is plotted on the bending moment diagram of Fig. 5-14*d*. Observing that the bending moment at both R_1 and R_3 is zero gives us two other points on the bending moment curve.

(5) The maximum positive moment in each span occurs at the section at which the shear passes through zero, as determined in step 3. Writing an expression for the maximum positive bending moment in the left span

$$M = (3.75 \times R_1) - (1000 \times 3.75 \times 3.75/2)$$

$$M = (3.75 \times 3750) - 7031 = 7031 \text{ ft-lb}$$

The same value occurs in the right span; both are plotted on the bending moment diagram.

(6) Two additional points on the moment curve that are readily determined are the inflection points (sometimes called points of contraflexure). Under the uniform loading of this example the positive moment curve in each span will be symmetrical about its apex (maximum value), which is located 3.75 ft from R_1 and R_3, respectively. Therefore the bending moment will be zero at a distance from R_1 and R_3 equal to 2×3.75 or 7.5 ft. We have now located seven points on the bending moment diagram and a smooth curve may be drawn between them, as shown in Fig. 5-14*d*.

Problem 5-7-A. The two spans of a continuous beam are each 10 ft in length. Each span carries a concentrated load of 4 kips at its center. Construct the shear and moment diagrams and identify the magnitude of the maximum shear and the maximum bending moment. (*Note:* See Fig. 5-13, Case 9.)

6

Shear Stresses in Beams

6-1 Vertical Shear

Shearing stresses in beams were discussed briefly in Sec. 2-4. The tendency for a beam to fail by vertical shear is illustrated in Fig. 2-3*b*. This type of shear is actually cross-grain shear, and failures of the kind indicated seldom occur in wood beams. It is generally unnecessary to investigate cross-grain or vertical unit shearing stresses. For simple beams the maximum vertical shear has the magnitude of the greater reaction, and methods for determining its magnitude were explained in Sec. 5-4.

6-2 Horizontal Shear in Rectangular Beams

Any beam subjected to vertical shear is also subjected to horizontal shearing forces. Horizontal shear is the tendency of one part of a beam to slide horizontally on an adjacent part, as illustrated in Fig. 2-3*c*. The horizontal unit shearing stresses are not uniformly distributed over the cross section of a beam but are greatest at the neutral surface. The formula used to compute the maximum horizontal unit shearing stress in an unchecked *rectangular* beam is

$$f_v = \frac{3}{2} \times \frac{V}{bd}$$

in which f_v = maximum unit horizontal shearing stress (psi),
V = total vertical end shear (lb),
b = breadth (width) of cross section (in.),
d = depth of cross section (in.).

This formula is derived in Sec. 6-4. It applies only to rectangular cross sections and yields generally conservative values since it indicates a greater stress than is normally present.

When investigating horizontal shear, it is customary to begin by using this formula. If f_v is found to be less than the allowable unit horizontal shearing stress for the species and grade (F_v in Table 3-1), the beam is adequate for shear. If f_v exceeds F_v, however, f_v is again computed using a modified value of the end shear V, as explained below.

Checks and shakes are present in nearly all structural lumber. Because of their presence, the upper and lower portions of a beam act partly as two beams and partly as a unit. The conservative results given by the formula for horizontal shearing stress are partly a consequence of this "two-beam" action. This fact has led to an empirical rule that permits the end shears (reactions) of a simple beam to be calculated by neglecting all loads within a distance equal to the depth of the beam from both supports. If, when using the modified value of the end shear, f_v still exceeds F_v, a beam of larger cross-sectional area may be selected.

Example 1. A simple beam with a span of 14 ft supports a uniformly distributed load of 800 lb per lin ft. A 10 × 14 beam of Douglas fir, Select Structural grade, is used. Is the beam safe with respect to horizontal shear?

Solution: (1) Referring to Table 3-1, we find that the allowable horizontal shearing stress is F_v = 85 psi. Table 4-1 shows that the dressed size of a 10 × 14 beam is 9.5 × 13.5 in.

(2) The total load on the beam is 800 × 14 = 11,200 lb, making each reaction and the maximum end shear equal to 11,200 ÷ 2 = 5600 lb. The unit horizontal shearing stress is

$$f_v = \frac{3V}{2bd} = \frac{3 \times 5600}{2 \times 9.5 \times 13.5} = 66 \text{ psi}$$

(3) We see that 66 psi is below the design value of 85 psi, so the beam is safe with respect to horizontal shear. However, if f_v had exceeded the allowable value, the method of neglecting a portion of the distributed load at each end of the beam to establish a modified shear could be tried. Although unnecessary in this example, Step 4 demonstrates how the method is employed.

(4) As already noted, all loads for a distance equal to the height of the beam from both supports are to be neglected. Because the beam is 13.5 in. deep (1.125 ft), the loaded length of span to be considered is (14 − 2.25) ft. The load per linear foot is 800 lb, making each modified reaction and end shear

$$V = \frac{800 \times (14 - 2.25)}{2} = 4700 \text{ lb}$$

and the unit horizontal shearing stress becomes

$$f_v = \frac{3V}{2bd} = \frac{3 \times 4700}{2 \times 9.5 \times 13.5} = 55 \text{ psi}$$

Of course, there is no reason why the shearing stress in a beam cannot be calculated by using modified end shears in the first place if it appears that horizontal shear may be critical in any given case.

Example 2. An 8 × 14 beam is used on a simple span of 16 ft and supports concentrated loads of 4000 lb at the quarter points of the span (Fig. 5-13, Case 5). The lumber used has an allowable horizontal shearing stress of 95 psi. Investigate the horizontal shear.

Solution: (1) For the given loading conditions, both reactions and the maximum end shear are equal to $3P/2$ or (3 × 4000) ÷ 2 = 6000 lb.

(2) The maximum horizontal shearing stress developed in the beam is

$$f_v = \frac{3V}{2bd} = \frac{3 \times 6000}{2 \times 7.5 \times 13.5} = 89 \text{ psi}$$

Since this stress is less than the allowable value of 95 psi, the 8 × 14 section is adequate with respect to horizontal shear.

6-3 Shear Investigation Summary

In the design of beams it is customary first to determine the size of the beam to withstand the bending stresses (Chapter 9). When the beam dimensions have thus been established, the beam is investigated for horizontal shear. The question to be resolved is whether the cross-sectional dimensions are adequate to result in a value of f_v that does not exceed the allowable horizontal unit shearing stress for the species and grade of lumber being used. Consequently the formula $f_v = 3V/2bd$ is tried. If f_v exceeds F_v, the beam is further investigated by making use of the modified reactions (end shears) illustrated in step 4 of Example 1, Sec. 6-2. If the unit shearing stress still exceeds the allowable value, another beam of greater cross-sectional area (that still satisfies the bending requirements) is selected and the shear test repeated. This procedure may be employed in solving the following problems.

Problem 6-3-A. A 10 × 12 [241 × 292 mm] beam of Southern pine, No. 1 Dense SR grade, is used on a simple span of 12 ft [3.6 m]. It supports a uniformly distributed load of 1200 lb per lin. ft [17.5 kN/m]. Is the beam large enough with respect to horizontal shear?

Problem 6-3-B. A 10 × 14 [241 × 343 mm] timber is used for a beam on a span of 15 ft [4.5 m] with a concentrated load of 9 kips [40 kN] located 5 ft [1.5 m] from the left end. Compute the maximum horizontal shearing stress.

Problem 6-3-C*. A 6 × 10 [14 × 241 mm] beam carries a uniformly distributed load of 6600 lb [29.4 kN] on a span of 10 ft [3m]. If the allowable horizontal shearing stress for the timber used is 85 psi [586 kPa], is the beam large enough with respect to horizontal shear?

6-4 General Formula for Shear

The formula we used in Sec. 6-2 to determine the maximum horizontal shearing stress in beams of rectangular cross section is

$$f_v = \frac{3V}{2bd}$$

Because timber beams commonly used have rectangular cross sections, this formula is generally appropriate. In recent years, however, important advances have been made in the fabrication of wood beams of various sizes and shapes. Glued-laminated beams are often employed and I-shapes and box beams are built up to comply with special requirements. To find the horizontal shearing stresses for beams in which the cross sections are not rectangular, we use the general formula for shear

$$f_v = \frac{VQ}{Ib}$$

In this expression

f_v = unit horizontal shearing stress at any specific point in the cross section of the beam (psi),
V = total vertical shear at the section selected (lb),
Q = statical moment with respect to the neutral axis of the area of the cross section above (or below) the point at which f_v is to be determined (in.^{3}) (a *statical moment* is an area multiplied by the distance of its centroid from a given axis),
I = moment of inertia of the beam cross section with respect to its neutral axis (in.^{4}),
b = width of beam at the point where v is to be computed (in.).

Consider the rectangular cross section of which b and d are the width and depth, respectively, as shown in Fig. 6-1a. Using the general formula, let us compute the maximum horizontal shearing stress *at the neutral axis X–X*.

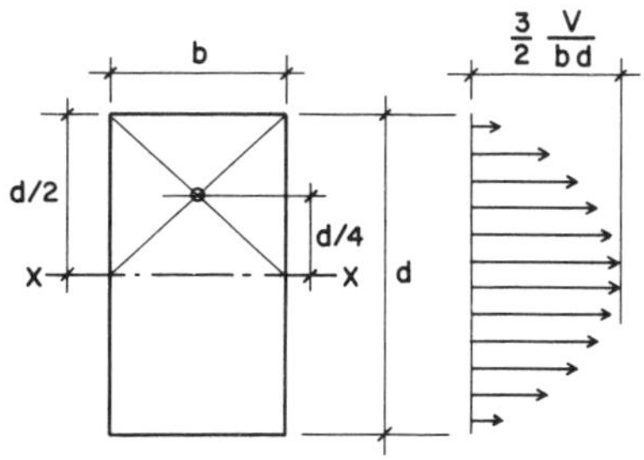

FIGURE 6-1.

The area above the neutral axis is ($b \times d/2$) and its centroid is at distance $d/4$ from the neutral axis. Then

$$Q = \left(b \times \frac{d}{2}\right) \times \frac{d}{4} = \frac{bd^2}{8}$$

From Sec. 4-3 we know that the moment of inertia of a rectangle is $I = bd^3/12$. Substituting in the general formula,

$$f_v = \frac{VQ}{Ib} = \frac{V \times bd^2/8}{bd^3/12 \times b} \quad \text{and} \quad f_v = \frac{3}{2} \times \frac{V}{bd}$$

which is the formula used to compute the maximum horizontal shearing unit stress for *rectangular* sections. The horizontal shearing stresses are not equally distributed over the area of the cross section. These stresses are indicated by the lengths of the arrows in Fig. 6-1*b*, the maximum stress being at the neutral axis.

Another use for the general formula for horizontal shear is to determine the stress at glue lines of built-up wood beams. The following is an example.

Example. Determine the unit shearing stress at the glue line of the box beam shown in Fig. 6-2 when the maximum vertical shear is 4000 lb.

Solution: (1) Determine the horizontal shearing stress in the glue joint under the upper 2 × 6 flange piece where it is joined to

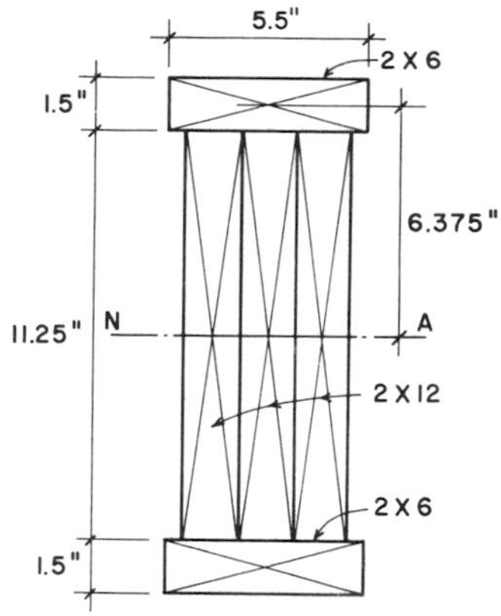

FIGURE 6-2.

the vertical 2 × 12 members. This is accomplished by using the general formula for shear

$$f_v = \frac{VQ}{Ib}$$

For easy reference the dressed dimensions of all members of the box section have been recorded in decimals in Fig. 6-2.

(2) The statical moment of the 2 × 6 top (or bottom) flange piece about the neutral axis of the entire built-up section is

$$Q = (1.5 \times 5.5) \times 6.375 = 52.6 \text{ in.}^3$$

(3) In the next step we calculate the moment of inertia of the entire cross section about its neutral axis. This consists of the moment of inertia of the three 2 × 12 pieces about their own centroidal axes (which coincide with the neutral axis of the built-up section) plus the I_0 of the two 2 × 6 pieces transferred to the neutral axis of the entire section. The I for one 2 × 12 member from Table 4-1 is 178 in.^{4}, which makes the combined I 3 × 178 = 534 in.^{4}.

To find I for the top 2 × 6 piece about the neutral axis we use the transfer formula from Sec. 4-4:

$$I = I_0 + Az^2$$

For I_0 we use the property listed for a 6 × 2 in Table 4-1, that is, 1.55 in.[4]. The area of the 2 × 6 is 8.25 in.[2] and the distance z, as shown in Fig. 6-2, is 6.375 in. Then

$$I = I_0 + Az^2 = 1.55 + 8.25(6.375)^2 = 337 \text{ in.}^4$$

For the two 2 × 6 pieces $I = 2 \times 337 = 674$ in.[4]. Then for the entire section $I_{NA} = 674 + 534 = 1208$ in.[4]

(4) The width of the beam at the glue line is 3 × 1.5 = 4.5 in. Substituting these values in the general formula for shear stress, we find

$$f_v = \frac{VQ}{Ib} = \frac{4000 \times 52.6}{1208 \times 4.5} = 38.7 \text{ psi}$$

Problem 6-4-A. Five pieces of lumber are glued together to form a box beam similar to that shown in Fig. 6.2. The upper and lower flanges are 2 × 8 pieces and the vertical members consist of two 3 × 10s and one 2 × 10. If the maximum vertical shear on this built-up beam is 6000 lb [27 kN], compute the unit shear stress at the glue line.

7

Bending Stresses in Beams

7-1 Resisting Moment

We learned in Sec. 5-5 that bending moment is a measure of the tendency of the external forces acting on a beam to deform it by bending. We will now consider the action within the beam that resists bending and is called the *resisting moment*.

Figure 7-1*a* shows a simple beam of rectangular cross section supporting a single concentrated load P. An enlarged sketch of the left portion of the beam, between the reaction and section X–X, is shown in Fig. 7-1*b*. From the discussions of bending moment in Chapter 5 we know that the reaction R_1 tends to cause a clockwise rotation about point A in the section under consideration; this we have defined as the bending moment at the section. In this simple beam the fibers in the upper part are in compression and those in the lower part are in tension. There is a horizontal plane separating the compressive and tensile stresses; as noted earlier, it is called the *neutral surface*. At this plane there are neither compressive nor tensile stresses with respect to bending.

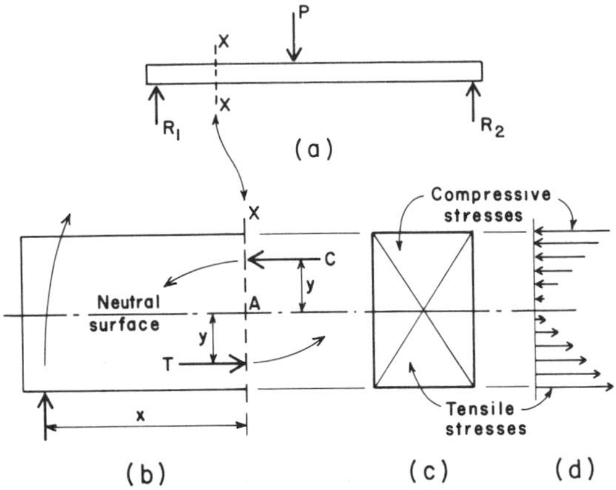

FIGURE 7-1. Development of bending stress.

The line in which the neutral surface intersects the beam cross section (Fig. 7-1*c*) is the *neutral axis* of the cross section.

Let C be the resultant of all the compressive stresses acting on the upper part of the cross section and T, the resultant of all the tensile stresses acting on the lower part. It is the sum of the moments of these stresses at the section that holds the beam in equilibrium; this is called the *resisting moment* and is equal in magnitude to the bending moment at the section. The bending moment at point A is $R_1 \times x$ and the resisting moment about the same point is $(C \times y) + (T \times y)$. The bending moment tends to cause a clockwise rotation and the resisting moment tends to cause a counterclockwise rotation. Because the beam is in equilibrium, we may write

$$R_1 \times x = (C \times y) + (T \times y)$$

or

$$\text{bending moment} = \text{resisting moment}$$

For any type of beam we can compute the maximum bending moment developed by the loading. If we wish to design a beam for this loading, we must select a member with a cross section of such shape, area, and material that it is capable of developing a resisting moment equal to the maximum bending moment; this is accomplished by use of the *flexure formula* (frequently called the *beam formula*) discussed in the following section.

7-2 Flexure Formula

The flexure formula is an expression for resisting moment that involves the size and shape of the beam cross section and the material of which the beam is made. It is used in the design of all homogeneous beams (i.e., beams made of one material only, such as steel, aluminum, or wood). The flexure formula is usually written as

$$M = f \times S$$

where the size and the shape of the cross section are represented by the section modulus S (Sec. 4-5) and the material of which the beam is made is represented by f, the unit stress on the fiber most remote from the neutral axis; this stress is called the *extreme fiber stress*. Although the reader will never need to derive the flexure formula, it will be used many times. The following brief discussion is presented to show the principles on which the formula is based.

In Fig. 7-1*b* the compressive and tensile stresses acting on the rectangular beam cross section were represented by their resultants C and T, respectively. However, neither the fibers in compression nor those in tension are stressed equally but vary from zero stress at the neutral axis to a maximum at the extreme top and bottom fibers. This pattern of stress distribution is shown in Fig. 7-2*b* with the extreme fiber stress (in both compression and tension) designated by f. Because the neutral axis passes through the centroid of the cross section (Fig. 7-2*a*), it is located midway between the upper and lower surfaces of the beam, and the distance of the most remote fibers from the neutral axis is $c = d/2$.

For the present let us consider the compressive stresses indicated by the hatched area in Fig. 7-2*a*; this area is $(b \times d/2)$.

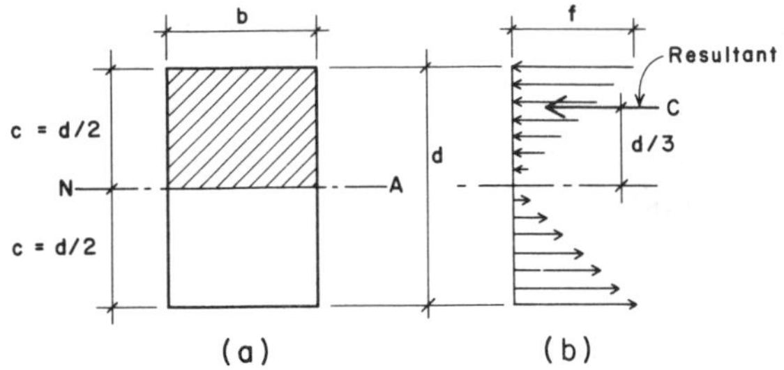

FIGURE 7-2.

Since the stress on the fibers at the neutral axis is zero, the *average* unit stress on the fibers in compression is $f/2$. This makes the total compressive stress (C) equal to $(b \times d/2) \times f/2$.

We know that the centroid of the compressive stress distribution triangle lies at two-thirds of the distance c (or $d/3$) above the neutral axis and that the resultant of all the compressive stresses passes through this point. Therefore the sum of the moments of all of the compressive stresses about the neutral axis is $(b \times d/2) \times f/2 \times d/3$ and we may write

$$M_C = b \times \frac{d}{2} \times \frac{f}{2} \times \frac{d}{3}$$

If we multiply this expression by 2 (to include the moments of all the tensile stresses below the neutral axis), we have an expression representing the sum of the moments of all the stresses in the cross section with respect to the neutral axis. The result is, of course, the resisting moment. Thus

$$M_R = M_C + M_T = 2 \times \left(b \times \frac{d}{2} \times \frac{f}{2} \times \frac{d}{3}\right)$$

or

$$M_R = f \times \frac{bd^2}{6}$$

Because the resisting moment is equal in magnitude to the bending moment, $M_R = M$ and the expression becomes

$$M = f \times \frac{bd^2}{6} \quad \text{or} \quad \frac{M}{f} = \frac{bd^2}{6}$$

Recall from the discussion in Sec. 4-5 that the quantity $bd^2/6$ is the section modulus of a rectangular cross section about an axis through the centroid parallel to the base. Consequently the last two equations are special forms of the flexure formula $M = f \times S$.

7-3 Application of the Flexure Formula

The expression $M = fS$ may be stated in three different forms, depending on the information desired. These are given here in a notation that makes a distinction between f as the *allowable* extreme fiber stress (F_b) and f as the *computed* extreme fiber stress (f_b). The terms extreme fiber stress and bending stress are often used interchangeably.

$$(1)\ M = F_b S \qquad (2)\ f_b = \frac{M}{S} \qquad (3)\ S = \frac{M}{F_b}$$

Form (1) gives the maximum potential resisting moment when the section modulus of the beam and the maximum allowable bending stress are known. Form (2) gives the computed bending stress when the maximum bending moment due to the loading is known, together with the section modulus of the beam. These two forms are used to investigate the adequacy of given beams.

Form (3) is the one used in design. It gives the *required* section modulus when the maximum bending moment and the allowable bending stress are known. When the required section modulus has been determined, a beam with an S equal to or greater than the computed value is selected from tables giving properties of the various structural shapes.

When using the beam formula, care must be exercised with respect to the units in which the terms are expressed. Bending stress values F_b and f_b are typically written in pounds per square inch (psi) or kips per square inch (ksi) [kPa or MPa in SI units]. S is usually given in in.^{3} or mm^3. Bending moments are usually computed in ft-lb or kip-ft, thus requiring a conversion from feet to inches [kN-m in SI units, requiring the usual attention to powers of 10].

Example. The maximum bending moment on a beam is 13,100 ft-lb [17.8 kN-m]. If the allowable bending stress for the wood is F_b = 1400 psi [9.65 MPa], determine the required beam section.

Solution: (1) The required section modulus is

$$S = \frac{M}{F_b} = \frac{13{,}100 \times 12}{1400} = 112.3 \text{ in.}^3 \; [1845 \times 10^3 \text{ mm}^3]$$

(2) From Table 4-1 select a 6 × 12 beam with S = 121.2 in.^{3}.

(3) Suppose that a limit for headroom under the beam requires a 10 in. deep beam. To find the required width for such a beam we may solve the formula for the section modulus for a rectangular shape as follows:

$$S = 112.3 = \frac{bd^2}{6}$$

$$b = \frac{6S}{d^2} = \frac{6 \times 112.3}{(9.5)^2} = 7.47 \text{ in.}$$

which indicates that an 8 × 10 (7.5 × 9.5 actual) section can be used. This could also have been accomplished by scanning of Table 4-1 for the narrowest 10 in. deep section with an S of at least 112.3 in.^{3}.

Problem 7-3-A. If the maximum bending moment on an 8 × 10 beam is 16 kip-ft [22 kN-m], what is the value of the maximum bending stress?

Problem 7-3-B. Find the required section with least area for a beam in which the maximum bending moment is 20 kip-ft [27 kN-m] if the allowable stress is 1600 psi [11 MPa].

7-4 Size Factor for Rectangular Beams

As the depth of a rectangular beam increases, there is a slight decrease in the unit bending strength. Consequently for solid-sawn beams deeper than 12 in. and thicker than 4 in. the allowable bending stress F_b must be reduced. This is accomplished by multiplying the tabular value of F_b by the appropriate *size factor* determined from the formula

$$C_F = \left(\frac{12}{d}\right)^{1/9}$$

in which C_F = size factor,
d = actual depth of beam (in.).

Values of C_F for a few different depths of solid-sawn beams are given in Table 7-1.

To determine the resisting moment of a beam deeper than 12 in. the size factor is inserted in the flexure formula to give the equation the form

$$M = C_F \times F_b \times S$$

For example, the allowable extreme fiber stress for beams and stringers of Douglas fir, Select Structural grade, is given in Table

TABLE 7-1. Size Factors for Solid-Sawn Beams[a]

Beam Depth		
(in.)	(mm)	C_F
13.5	343	0.987
15.5	394	0.972
17.5	445	0.959
19.5	495	0.947
21.5	546	0.937
23.5	597	0.928

[a] For U.S. units $C_F = (12/d)^{1/9}$. For SI units $C_F = (300/d)^{1/9}$.

3-1 as $F_b = 1600$ psi. For a nominal 8 × 16 beam, however, a size factor of 0.972 must be applied. Noting from Table 4-1 that the section modulus of a nominal 8 × 16 cross section is 300 in.^{3}, we find that the resisting moment of such a beam is

$$M = C_F F_b S = 0.972 \times 1600 \times 300 = 467{,}000 \text{ in.-lb.}$$

8

Deflection of Beams

8-1 General

The deformation that accompanies bending of a beam is called *deflection*. Figure 8-1 illustrates the deflection of a simple beam; we are principally concerned with its maximum value which, in this instance, occurs at the center of the span. Some degree of deflection exists in all beams, and the designer must see that the deflection does not exceed certain prescribed limits. It is important to understand that a beam may be adequate to support the imposed loading without exceeding the allowable bending stress, but at the same time the curvature may be so great that cracks appear in suspended plaster ceilings, water collects in low spots on flat roofs, or the general lack of rigidity results in an excessively springy floor. In other words, a beam should be designed for strength in bending and also for *stiffness*.

8-2 Allowable Deflection

The allowable limit for the deflection of beams in floor construction that supports plaster ceilings or partitions is generally taken to be $\frac{1}{360}$ of the span. For such construction an initial deflection

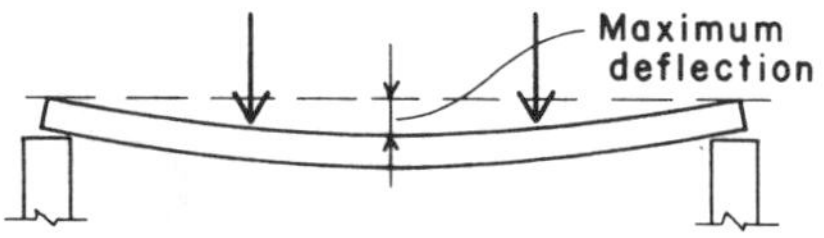

FIGURE 8-1. Beam deflection.

due to the dead load (the weight of the materials of construction) occurs before the plaster is applied. It is for this reason that the dead load is sometimes omitted when computing deflection. On the other hand, the sag of beams under dead load can be unattractive both visually and psychologically, and many designers frequently include both dead and live loads when determining deflection. Some authorities consider the allowable deflection limit for beams that do not support plastering to be $\frac{1}{240}$ of the span length.

When the types of occupancy that may occur in a building throughout its life are unknown, the more restrictive limitation ($\frac{1}{360}$) is appropriate. When the future use is obvious, however, advantage may be taken of the economy provided by a greater permissible deflection. In practice, of course, the local building code should always be consulted for specific provisions governing deflection.

The maximum permissible deflection for a beam is sometimes expressed as an absolute dimension rather than a proportion of the span. Such a dimension is usually established from known construction clearance requirements; for example, if a beam deflects more than a specified amount, it is known that damage will be caused to adjacent materials such as window glass and frames, door frames, and nonbearing partitions.

8-3 Computation of Deflections

Although considerations for deflection can be serious, the precise determination of deflections is an impractical and essentially unattainable goal. Some reasons for this are as follows:

Determination of loads always involves some degree of approximation.

The modulus of elasticity of any individual piece of wood is always an approximate value.

Various restraints on structural deformation always exist due to load sharing, connection resistances, stiffening due to nonstructural elements of the construction, and so on.

As a result of this situation—and owing to the often quite laborious nature of deflection computations—it is usually adequate for design purposes to use various shortcut approximate techniques for finding deflections.

For typical, frequently occurring load and span situations it is possible to derive formulas for beam deflection. For most of the cases presented in Fig. 5-13 the appropriate formula is given for determining the maximum deflection. In most design problems, it is only the maximum value of the deflection that is of concern. Where these loading situations occur (constituting the vast majority of all beams used) these formulas may be used.

(Note that the traditional symbol for deflection in structural work is the Greek capital letter Δ—delta. However, some references also use an English capital *D*.)

Deflections in wood structures tend to be most critical for rafters and joists where span–to–depth ratios are often pushed to the limit. Maximum permitted spans for particular arrangements of rafters and joists are often limited by consideration of deflection. Since rafters and joists are usually of a simple span form with uniformly distributed loading (Case 2 in Fig. 5-13), the deflection takes the form of the equation

$$\Delta = \frac{5\,WL^3}{384\,EI}$$

By substitution of relations involving W, L, and flexural stress that accompanies the deflection, the deflection can be described as

$$\Delta = \frac{5\,L^2 f_b}{24\,Ed}$$

If average values of f_b = 1500 psi and E = 1,500,000 psi are substituted in this equation, it reduces to the form

$$\Delta = \frac{0.03\ L^2}{d}$$

in which Δ = the deflection in inches,
L = the beam span in feet,
d = the beam depth in inches.

Figure 8-2 contains plots of this formula for various values of d that represent standard lumber dimensions. The curves in Fig. 8-2 are labeled by nominal dimensions, but the computations for the plots were done with the true dimensions as given in Table 4-1. For reference, lines are shown corresponding to ratios of deflection of $L/240$ and $L/360$. Also shown for reference is the line that represents a span–to–depth ratio of 25 to 1, which is an approximate practical limit.

For beams with values of f_b and E other than those used for the plots in Fig. 8-2, actual deflections can be obtained as follows:

$$\text{True } \Delta = \frac{\text{true } f_b}{1500} \times \frac{1{,}500{,}000}{\text{true } E \text{ in psi}} \times \Delta \text{ value from graph}$$

Beams with unsymmetrical or complex loadings often present difficult situations for deflection computation. If the beams are of a simple span type, an approximate maximum deflection can be obtained in many cases by using an equivalent uniform load derived by equating the actual maximum moment in the beam to one produced by a hypothetical uniform load. This hypothetical "W" can then be used with the equation for Case 2 in Fig. 5-13 or with the graphs in Fig. 8-2 for an approximate determination of the deflection.

The following examples illustrate some typical problems involving computation of deflections and illustrate the techniques just discussed.

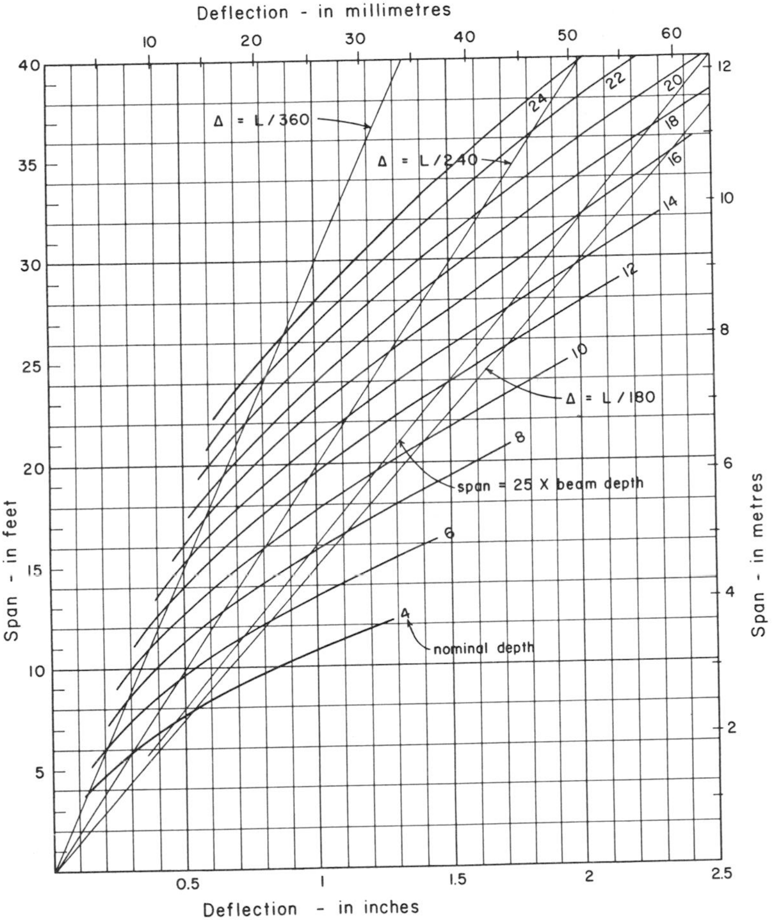

FIGURE 8-2. Deflection of wood beams. Assumed conditions: maximum f_b = 1500 psi, E = 1,500,000 psi.

Example 1. An 8 × 12 wood beam with E = 1,600,000 psi is used to carry a total uniformly distributed load of 10,000 lb on a simple span of 16 ft. Find the maximum deflection.

Solution: Using the formula for this loading (Case 2 in Fig. 5-13) and the value of I = 950 in.[4] for the section obtained from Table

4-1, we compute

$$\Delta = \frac{5\,WL^3}{384\,EI} = \frac{5(10{,}000)(16 \times 12)^3}{384(1{,}600{,}000)(950)} = 0.61 \text{ in.}$$

Or, using the graphs in Fig. 8-2

$$M = \frac{WL}{8} = \frac{(10{,}000)(16)}{8} = 20{,}000 \text{ ft-lb}$$

$$f_b = \frac{M}{S} = \frac{(20{,}000 \times 12)}{165} = 1455 \text{ psi}$$

From Fig. 8-2, Δ = approximately 0.66 in. Then

$$\text{True } \Delta = \frac{1455}{1500} \times \frac{1{,}500{,}000}{1{,}600{,}000} \times 0.66 = 0.60 \text{ in.}$$

which shows reasonable accuracy in comparison with the computation.

Example 2. Find the maximum deflection for the beam in Example 3 of Sec. 5-5 if the section is a 6 × 10 with E = 1,400,000 psi. (See Fig. 5-11.)

Solution: Because no formulas are given in Fig. 5-13 for this loading, the procedure used will be that of finding an approximate deflection using the equivalent uniform load method.

(1) Using the maximum value for moment as given in Fig. 5-11, we determine if

$$M = 6600 = \frac{WL}{8}, \quad W = \frac{8M}{L} = \frac{8(6600)}{18} = 2933 \text{ lb}$$

(2) With this equivalent uniform load we determine

$$\Delta = \frac{5\,WL^3}{384\,EI} = \frac{5(2933)(18 \times 12)^3}{384(1{,}400{,}000)(393)} = 0.70 \text{ in.}$$

Or (3) using the graphs in Fig. 8-2, we first find

$$f_b = \frac{M}{S} = \frac{6600 \times 12}{83} = 954 \text{ psi}$$

From Fig. 8-2, Δ = approximately 1.03 in. Then

$$\text{True } \Delta = \frac{954}{1500} \times \frac{1{,}500{,}000}{1{,}400{,}000} \times 1.03 = 0.70 \text{ in.}$$

Problem 8-3-A*. A 4 × 14 beam with E = 1,800,000 psi is used to carry a uniformly distributed load with a total value of 8000 lb on a simple span of 20 ft. Find the maximum deflection. Work the problem using the deflection formula and also using Fig. 8-2 and compare the results.

Problem 8-3-B. A wood beam is loaded as shown for Case 5 in Fig. 5-13. The beam is a 10 × 16 with E = 1,300,000 psi. The value of the concentrated load P is 4 kips and the beam span is 24 ft. Find (1) the maximum deflection by using the formula given in Fig. 5-13; (2) an approximate value for the deflection using the equivalent uniform load and Fig. 8-2.

9

Design of Beams

9-1 General

As noted earlier, a beam is a structural member subjected to transverse loads. Definitions for several types of beams are given in Sec. 5-1 and illustrated in Fig. 5-1. The comparatively small, closely spaced beams that directly support the subfloor in wood frame construction are called *joists*. Inclined joists used to support sloping roofs are called *rafters*. In this chapter we consider the general design of rectangular beams with nominal cross-sectional dimensions of 5 × 8 in. and larger; that is, members consisting of structural lumber that falls under the classification *beams and stringers* (Sec. 1-6). The design of joists and rafters is presented in Chapter 10.

So far in our discussion of beams we have treated reactions as concentrated forces. This, of course, is only an approximation, as can be seen from the case of a beam resting on a masonry wall (Fig. 9-1). The beam extends past the face of the wall a certain distance and the reaction is distributed over the area of contact between the beam and the wall. This area, known as the *bearing area,* is usually sufficiently small so that no appreciable error is introduced by considering the reaction to act at the center of the

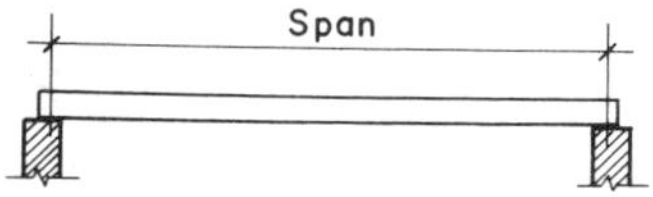

FIGURE 9-1.

bearing area. Therefore the span of a simple beam may be taken as the clear distance between faces of supports plus half the required length of bearing at each end.

9-2 Bearing Length

The end bearings of wood beams must have ample dimensions so that the compressive stresses perpendicular to grain do not exceed the allowable values of $F_{c\perp}$ given in Table 3-1. The allowable stresses in the table apply to bearings of any length at the ends of beams and to all bearings 6 in. or more in length at any other location.

For bearings less than 6 in. in length and not nearer than 3 in. to the end of a member the NDS provides that the allowable stress in compression perpendicular to grain may be increased by the factor $(l_b + 0.375)/l_b$ in which l_b is the length of bearing measured along the grain of the wood. This expression yields multiplying factors tabulated for the bearing lengths indicated:

Bearing length (in.)	$\frac{1}{2}$	1	$1\frac{1}{2}$	2	3	4	6 or more
Factor	1.75	1.38	1.25	1.19	1.13	1.10	1.0

Example. An 8 × 14 Southern pine beam, No. 1 SR grade, has a bearing length of 6 in. [150 mm] at its supports. If the end reaction is 8000 lb [36 kN], is the beam adequate with respect to compressive stress perpendicular to grain?

Solution: (1) Table 4-1 shows that the dressed width of this beam is 7.5 in. [190 mm], which makes the contact-bearing area $7.5 \times 6 = 45$ in.^{2} [28,500 mm^2].

(2) The bearing stress developed is equal to the reaction divided by the bearing area, or $8000/45 = 178$ psi [1.26 MPa].

(3) Referring to Table 3-1, we find that the allowable compressive stress perpendicular to grain is $F_{c\perp} = 375$ psi [2.60 MPa]. Because this value is greater than the developed stress, the bearing length is more than adequate.

Problem 9-2-A. A 6 × 10 beam of Douglas fir–larch, No. 1 grade, has 4 in. [100 mm] of end-bearing length and an end reaction force of 6 kips [27 kN]. Find (1) if the end bearing provided is adequate and (2) what minimum end-bearing length is required.

9-3 Design Procedure

The complete design of a wood beam is accomplished in five steps, some of which may be abridged as the designer gains experience; for example, many designers select a beam to satisfy step 2 that they know from experience will meet the requirements for shear and/or deflection. Experience in structural work will also enable the designer to estimate fairly accurately the weight of the beam so that an allowance for beam weight may be included under step 1. Weights per linear foot for structural lumber are given in Table 4-1, based on an assumed average weight of 35 lb per cu ft. The five steps of the design procedure follow.

Step 1. Compute the loads the beam will be required to support and make a dimensioned sketch (beam diagram) to show the loads and their locations. Find the reactions.

Step 2. Determine the maximum bending moment and compute the required section modulus from the flexure formula $S = M/F_b$, as explained in Sec. 7-3. A beam cross section with an adequate section modulus may then be selected from Table 4-1. Obviously many different sizes may meet this requirement, but the most practical have widths ranging from one-half to one-third the depth. Members that are relatively narrow tend to bend sidewise unless properly braced (Sec. 9-5).

Step 3. Investigate the beam selected in step 2 for horizontal shear following the procedure given in Sec. 6-3. If necessary, increase the dimensions of the beam. Allowable horizontal shearing stresses are given in Table 3-1.

Step 4. Investigate the beam for deflection to check that the computed deflection does not exceed the prescribed limit. Formulas for deflection under different loadings are given in Fig. 5-13 and are used as explained in Sec. 8-3.

Step 5. When a cross section that satisfies these requirements has been determined, the bearing length is established. As explained in Sec. 9-2, the bearing area must be large enough so that the allowable compressive stress perpendicular to grain will not be exceeded.

9-4 Examples Illustrating Beam Design

The following examples illustrate application of the foregoing procedure to the design of wood beams under different loading patterns.

Example 1. A simple beam has a span of 14 ft [4.2 m] and carries a load of 7200 lb [32 kN] uniformly distributed over the span length. Design the beam using Southern pine, No. 1 SR grade lumber. Deflection is limited to 0.5 in. [13 mm].

Solution: (1) The loading (except for the beam weight) is shown in Fig. 9-2, and applicable design information from Table 3-1 is recorded as follows: F_b = 1350 psi [9.3 MPa], F_v = 110 psi [0.76 MPa], $F_{c\perp}$ = 375 psi [2.60 MPa], E = 1,500,000 psi [10.3 GPa]. Because the loading is symmetrical, $R_1 = R_2$ = 7200/2 = 3600 lb [16 kN].

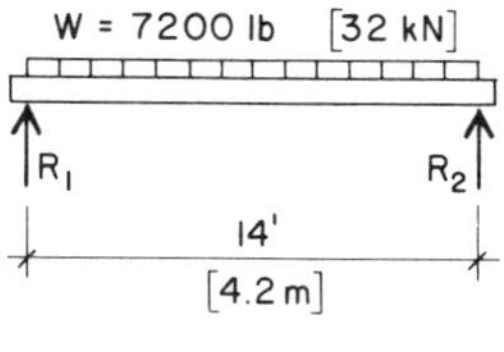

FIGURE 9-2.

(2) The maximum bending moment for this loading is

$$M = \frac{WL}{8} = \frac{(7200)(14)}{8} = 12{,}600 \text{ ft-lb } (16.8 \text{ kN-m}]$$

and

$$S = \frac{M}{F_b} = \frac{12{,}600 \times 12}{1350} = 112 \text{ in.}^3 \ [1806 \times 10^3 \text{ mm}^3]$$

Referring to Table 4-1, we find that a 6 × 12 beam has a section modulus of 121.3 (rounded tabular value) and weighs approximately 15.4 lb per lin ft, making the beam weight 15.4 × 14 = 216 lb. If the computations are repeated with a revised load of 7200 + 246 = 7416 lb [33.1 kN], the new bending moment is 13,000 ft-lb [17.4 kN-m] and the new required section modulus is 116 in.^{3} [1871 × 10^3 mm^3]. This is still less than that of the 6 × 12, so the chosen size is adequate.

(3) The maximum vertical end shear is equal to one of the reactions, or $R_1 = R_2 = V = 7416/2 = 3708$ lb [16.5 kN]. The horizontal shearing stress developed by the loading is

$$f_v = \frac{3}{2}\frac{V}{bd} = \frac{3}{2}\frac{3708}{63.25} = 87.9 \text{ psi } [0.608 \text{ MPa}]$$

This value is acceptable as it is lower than the allowable F_v.

(4) Investigating for deflection, we note that Case 2 of Fig. 5-13 applies to the conditions of this example. Table 4-1 shows that the moment of inertia of a 6 × 12 is 697.1 in.^{4}. Substituting this value and other design data in the deflection equation,

$$\Delta = \frac{5\,WL^3}{384\,EI} = \frac{5(7416)(14 \times 12)^3}{384(1{,}500{,}000)(697.1)} = 0.44 \text{ in. } [11 \text{ mm}]$$

Because the computed deflection is less than the specified limit, the 6 × 12 meets all the requirements and therefore is adopted.

(5) The minimum area required for the end bearing of the beam is equal to the reaction divided by F_c, or 3708/375 = 9.89 in.^{2} [6370 mm^2]. Table 4-1 shows that the true width of the 6 × 12 is 5.5 in. [140 mm], thus making the required bearing length 9.89/5.5 = 1.80 in. [45.5 mm]. As a matter of practical construction a beam of this size would usually be given a 4 in. minimum bearing length so that bearing is not a critical concern for this example.

Example 2. A simple beam has a span of 15 ft [4.5 m] with concentrated loads of 3200 lb [14 kN] placed at each third point of the span. There is also a uniformly distributed load of 200 lb/ft [2.9 kN/m] (including an allowance for the beam weight) extending over the entire span. Deflection is limited to $\frac{1}{360}$ of the span. Design the beam using Douglas fir–larch, No. 1 grade lumber.

Solution: (1) The load diagram is drawn as shown in Fig. 9-3, and the applicable design information from Table 3-1 is the following: F_b = 1300 psi [9.0 MPa]; F_v = 85 psi [0.59 MPa]; $F_{c\perp}$ = 625 psi [4.3 MPa]; and E = 1,600,000 psi [11 GPa]. The total symmetrical load on the beam is 3200 + 3200 + (200 × 15) = 9400 lb and $R_1 = R_2$ = 9400/2 = 4700 lb [20.525 kN].

(2) Referring to Fig. 5-13, we see that the maximum bending moment for both Case 2 and Case 3 occurs at the center of the span. The moments therefore may be added directly. Then

$$M = \frac{PL}{3} + \frac{WL}{8} = \frac{(3200)(15)}{3} + \frac{(3000)(15)}{8}$$

$$= 16{,}000 + 5625 = 21{,}625 \text{ ft-lb } [28.34 \text{ kN-m}]$$

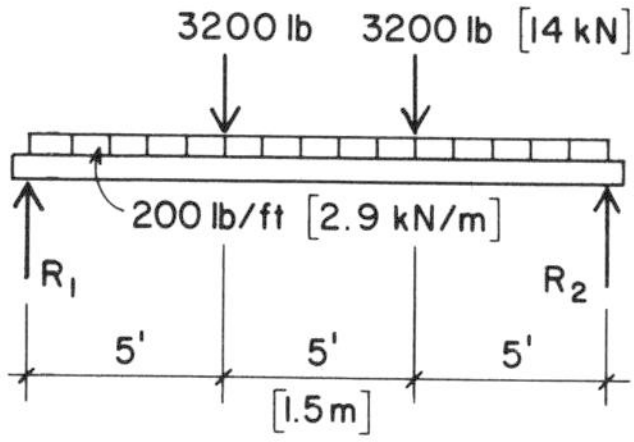

FIGURE 9-3.

and the required section modulus is

$$S = \frac{M}{F_b} = \frac{(21{,}625 \times 12)}{1300} = 199.6 \text{ in.}^3 \; [3150 \times 10^3 \text{ mm}^3]$$

Referring to Table 4-1, we find that $S = 209.4$ in.^{3} for a 10×12 beam and $S = 227.8$ for an 8×14 beam. If the 8×14 is selected, the allowable bending stress must be reduced by the size factor, as explained in Sec. 7-4. For the 14 in. deep beam the factor is 0.987, so the adjusted value of S is $199.6/0.987 = 202.2$ in.^{3} [3191×10^3 mm^3]. Because this value is still smaller than the actual value for the 8×14 beam, the choice is still valid.

(3) The maximum vertical shear is 4700 lb [20.525 kN], making the maximum horizontal shearing stress equal to

$$f_v = \frac{3}{2}\frac{V}{bd} = \frac{3}{2}\frac{4700}{101.3} = 69.6 \text{ psi } [0.471 \text{ MPa}]$$

Because this is less than F_v, the beam is adequate for horizontal shear.

(4) For this beam the maximum deflection occurs at the center of the span for both loadings, and can thus be found by combining the formulas for Case 2 and Case 3 from Fig. 5-13. The moment of inertia for the 8×14 from Table 4-1 is 1538 in.^{4}. Then

$$\Delta = \frac{23PL^3}{648EI} + \frac{5WL^3}{384EI} = \frac{23(3200)(15 \times 12)^3}{648(1{,}600{,}000)(1538)} + \frac{5(3000)(15 \times 12)^3}{384(1{,}600{,}000)(1538)}$$

$$= 0.269 + 0.093 = 0.362 \text{ in. } [6.8 + 2.4 = 9.2 \text{ mm}]$$

This is less than the allowable of $(15 \times 12)/360 = 0.5$ in. [13 mm] and so the beam is adequate for deflection.

(5) The required bearing area at each support is equal to the reaction divided by $F_{c\perp}$, or $4700/625 = 7.52$ in.^{2} [4850 mm^2]. With the beam width of 7.5 in., the minimum required bearing length is

7.52/7.5 = 1.0 in. Here, as in Example 1, the minimum bearing would probably be 4 in. and thus this factor is not critical for the beam design.

Example 3. A simple beam has a span of 12 ft [3.6 m] and supports a concentrated load of 6 kips [27 kN] at 4 ft [1.2 m] from the left support. In addition there is a uniformly distributed load of 1 k/ft [15 kN/m] (including an allowance for the beam weight) over the entire span. If deflection is limited to $\frac{1}{360}$ of the span, design the beam using the following values: F_b = 1600 psi [11 MPa]; F_v = 105 psi [0.72 MPa]; $F_{c\perp}$ = 375 psi [2.60 MPa]; and E = 1,600,000 psi [11 GPa].

Solution: (1) the beam diagram showing the loading is given in Fig. 9-4*a* and reactions are as shown.

(2) Maximum moment occurs where the shear diagram passes through zero, and may be computed as follows:

$$M = (10{,}000 \times 4) - (1000 \times 4 \times 2) = 32{,}000 \text{ ft-lb } [43.2 \text{ kN-m}]$$

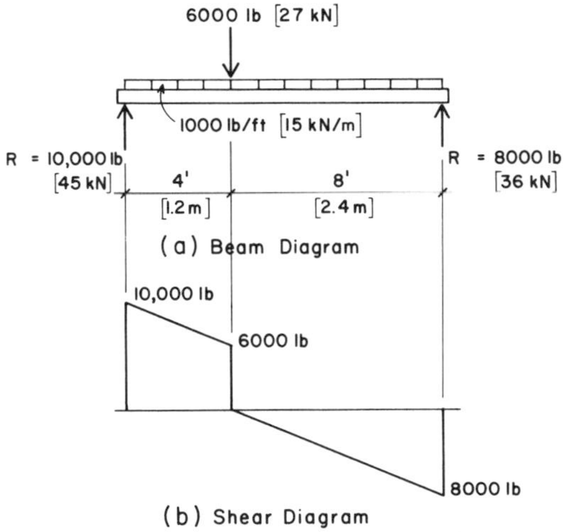

FIGURE 9-4.

Then

$$S = \frac{M}{F_b} = \frac{(32{,}000 \times 12)}{1600} = 240 \text{ in.}^3 \ [3930 \times 10^3 \text{ mm}^3]$$

From Table 4-1, a 10 × 14 has S = 288.6 in.³ This section is selected as a trial beam.

(3) Using the maximum shear from Fig. 9-4*b*, we compute

$$f_v = \frac{3}{2}\frac{V}{bd} = \frac{3}{2}\frac{10{,}000}{128} = 117 \text{ psi } [0.81 \text{ MPa}]$$

Because this exceeds F_v, we try again, and select the next widest section, that is, a 12 ×14. Then

$$f_v = \frac{3}{2}\frac{10{,}000}{155} = 97 \text{ psi } [0.67 \text{ MPa}]$$

which is acceptable. (Note that other selections are also possible, such as a 10 × 16. In a real situation various design factors not considered here may affect the choice between viable alternatives.)

(4) Since the loading on this beam is not one of the standard loadings listed in Fig. 5-13, we use an approximate method with an equivalent uniform load, as discussed in Sec. 8.3. The value for this hypothetical load is derived from the maximum bending moment; thus

$$W = \frac{8M}{L} = \frac{8(32{,}000)}{12} = 21{,}300 \text{ lb } [96 \text{ kN}]$$

Using the value for the moment of inertia of the 12 × 14 from Table 4-1, and substituting in the formula for the uniformly loaded simple beam, we compute

$$\Delta = \frac{5WL^3}{384EI} = \frac{5(21{,}300)(12 \times 12)^3}{384(1{,}600{,}000)(2358)} = 0.22 \text{ in. } [5.6 \text{ mm}]$$

The allowable deflection is (12 × 12)/360 = 0.4 in. [10 mm]. The beam is thus acceptable for deflection.

(5) Assuming a minimum bearing of 4 in. [100 mm], and using the value for the maximum reaction, we compute the end-bearing stress as

$$f_p = \frac{10{,}000}{4 \times 11.5} = 217 \text{ psi } [1.5 \text{ MPa}]$$

Because this is less than $F_{c\perp}$, bearing is not critical.

Problem 9-4-A*. A simple beam with a span of 15 ft [4.5 m] carries a uniformly distributed load of 9000 lb [40 kN] in addition to its own weight. Deflection is limited to 0.625 in. Design the beam using Douglas fir–larch, Select Structural grade.

Problem 9-4-B. A simple beam with a span of 13 ft [3.9 m] has concentrated loads of 9000 lb [40 kN] at each third point of the span. Deflection is limited to $\frac{1}{360}$ of the span. Design the beam using Southern pine, No. 1 SR grade.

9-5 Lateral Support for Beams

Specifications provide for the adjustment of bending capacity or allowable stress when a beam is vulnerable to a compression buckling condition. When beams are incorporated into framing systems, they are often provided with lateral bracing that is adequate to resist buckling effects, thus allowing use of the full value for flexural stress. Requirements for the type of bracing required to prevent both lateral and torsional buckling are given in NDS, Sec. 4.4.1, and are summarized in Table 9-1. If details of the construction do not provide for adequate bracing, then the rules of NDS, Sec. 3.3.3, must be used to reduce the usable bending capacity.

9-6 Unsymmetrical Bending

Beams in flat-spanning floor or roof systems are ordinarily positioned so that the loads and the plane of bending moment are perpendicular to one of the principal centroidal axes of the beam

TABLE 9-1. Lateral Support Requirements for Wood Beams[a]

Ratio of Depth to Thickness[b]	Required Conditions
2 to 1 or less	No support required.
3 to 1, 4 to 1	Ends held in position to resist rotation.
5 to 1	One edge held in position for entire span.
6 to 1	Bridging or blocking at maximum spacing of 8 ft; or both edges held in position for entire span; or one edge held in position for entire span (compression edge) and both ends held against rotation.
7 to 1	Both edges held in position for entire span.

[a] Adapted from data in Sec. 4.4.1 of *National Design Specification for Wood Construction*, 1986 edition, with permission of the publishers, National Forest Products Association.
[b] Ratio of nominal dimensions for standard sections.

section (x-axis as shown in Table 4-1). In this case the bending stresses are symmetrically distributed and the principal axis about which bending occurs is also the neutral stress axis.

There are various situations in which a structural member is subjected to bending in a manner that results in simultaneous bending about both principal axes of the beam section. If the member is braced against torsion and buckling, the result may simply be a case of what is called *biaxial bending* or *unsymmetrical bending*. Figure 9-5 shows a situation in which a roof beam is used for a sloping roof, that is, spanning between trusses or other beams that generate a sloping roof profile. With respect to gravity loads, the beam will experience bending in a plane that is rotated with respect to its major axes, resulting in components of bending about both its axes, as shown in Fig. 9-5*b*. The bending stresses about the two axes of the beam section produce the following stresses:

$$f_x = \frac{M_x}{S_x} \quad \text{and} \quad f_y = \frac{M_y}{S_y}$$

These are maximum stresses that occur at the edges of the section. Their distribution is of a form such as that shown in Fig.

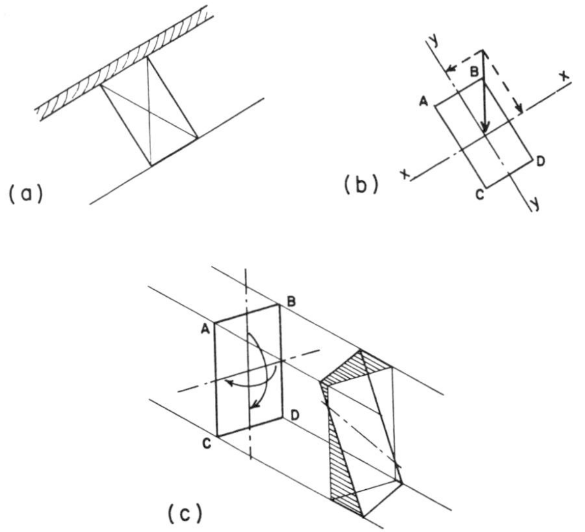

FIGURE 9-5. Development of unsymmetrical bending.

9-5*c*, which can be described by determining the values for the stresses at the four corners of the section. Noting the senses of the moments with respect to the two axes, and using plus for compressive stress and minus for tensile stress, the net stresses at the corners are as follows:

at *A*: $-f_x + f_y$
at *B*: $-f_x - f_y$
at *C*: $+f_x + f_y$
at *D*: $+f_x - f_y$

This is a somewhat idealized condition that ignores problems of torsional effects and potential lateral or torsional buckling, and is valid only for sections with biaxial symmetry (such as a rectangle or I-shape). If a member is used in the situation shown in Fig. 9-5, it should preferably be one with low susceptibility to these actions, such as the almost square section shown, or the construction should be carefully developed so as to provide bracing to prevent actions other than the simple bending illustrated in Fig. 9-5.

Beams that occur in exterior walls, although having an orientation that is vertical, may experience biaxial bending under a combination of gravity and lateral loads. Columns subjected to bending present another case for possible multiple bending conditions.

Example. Figure 9-6*a* shows the use of a beam in a sloping roof. The beam section is rotated to a position that corresponds to the roof slope of 30°, and consists of a wood member with a nominal size of 8 × 10. Find the net bending stress condition for the beam if the gravity load generates a moment of 10 kip-ft [14 kN-m] in a vertical plane.

Solution: From Table 4-1 we find the properties of the beam to be S_x = 112.8 in.^{3} [1.85 × 10^6 mm^3] and S_y = 89.1 in.^{3} [1.46 × 10^6 mm^3]. The components of the moment with respect to the major and minor axes of the section are:

$$M_x = 10 \cos 30 = 10(0.866) = 8.66 \text{ kip-ft } [12.12 \text{ kN-m}]$$

$$M_y = 10 \sin 30 = 10(0.5) = 5 \text{ kip-ft } [7 \text{ kN-m}]$$

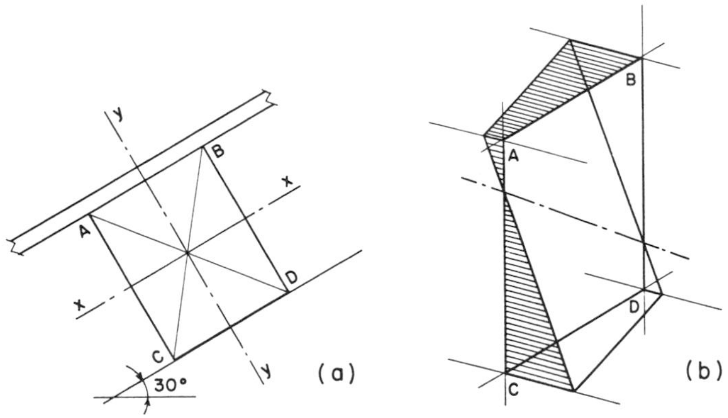

FIGURE 9-6.

and the corresponding maximum stresses are:

$$f_x = \frac{M_x}{S_x} = \frac{8.66 \times 12}{112.8} = 0.921 \text{ ksi } [6.55 \text{ MPa}]$$

$$f_y = \frac{M_y}{S_y} = \frac{5 \times 12}{89.1} = 0.673 \text{ ksi } [4.79 \text{ MPa}]$$

The net stress conditions at the four corners of the section—lettered A through D in Fig. 9-6*a*—are then determined as follows, using plus for compressive stress and minus for tensile stress:

at A: $-0.921 + 0.673 = -0.248$ ksi [-1.76 MPa]
at B: $-0.921 - 0.673 = -1.594$ ksi [-11.34 MPa]
at C: $+0.921 + 0.673 = +1.594$ ksi [$+1.76$ MPa]
at D: $+0.921 - 0.673 = +0.248$ ksi [$+11.34$ MPa]

The form of the stress distribution is as shown in Fig. 9-6*b*.

Problem 9-6-A. A wood beam consisting of a 6 × 10 of Douglas fir–larch, Dense No. 1 grade, is used as a beam in a sloping roof, as shown in Fig. 9-6. Assuming no stress modifications apply and that torsion and buckling are prevented, is the beam adequate for a bending moment of 8 kip-ft [10.8 kN-m] in a vertical plane with a roof slope of 25°?

10

Joists and Rafters

10-1 Joist Floors

Joists are the comparatively small, closely spaced beams that support the subfloor in wood construction. The nominal sizes commonly used for moderate loads are 2 × 8, 2 × 10, and 2 × 12. Although the strength of the structural subfloor is a factor, the spacing of joists is determined principally by the stock lengths of metal lath and gypsum board lath used as a base for plaster ceilings and the standard lengths of plywood, drywall, and other sheet materials that are applied directly to the joists. The standard width of lath material is 48 in. and the standard length of sheet material is 96 in.; thus common spacings for joists are 16 in. and 24 in. on centers (o.c.). When the floor load or length of span is excessive, a 12 in. spacing may be necessary.

Typical wood joist floor construction is illustrated in Fig. 10-1. The subfloor may consist of plywood sheets or of nominal 1 in. boards, generally laid diagonally with the joists to stiffen the construction. Wood finish flooring is indicated in the figure. A drywall ceiling is shown here, applied directly to the undersides of the joists. The purpose of *bridging* is twofold. It serves to prevent buckling by maintaining the joists in a vertical position and helps to distribute concentrated loads to adjacent joists. This latter function is partly the basis for the higher values of F_b given in Table 3-1 under "repetitive-member uses" (see Sec. 3-2). Material for cross bridging consists of fabricated metal straps or wood pieces cut from 2 × 3 or 2 × 4 stock. Bridging should be installed at intervals of about 8 ft along the span length.

An alternate method for bracing joists is by use of solid *blocking,* which consists of short pieces of joist fitted between the spanning joists in a continuous row. Blocking may be required for the development of the floor deck as a horizontal diaphragm, as discussed in Chapter 19. Blocking is also usually required under any walls that sit on the joist floor and are perpendicular to the joists.

10-2 Design of Joists

Because joists are really small simple beams, the design procedure for beams given in Sec. 9-3 may be applied to the design of floor joists. In practice, however, the design of joists that carry uniformly distributed loads (by far the most common loading) is normally accomplished by the use of tables that list maximum safe spans for different combinations of joist size, spacing, and loading. The use of these tables is explained in Sec. 10-3. This section presents two examples of joist design that use the usual procedures followed for beams. Example 1 is for a uniform loading over the entire span and is given to provide background for the subsequent use of span tables. Example 2 covers a special case in which the joists support a partition in addition to the uniformly distributed floor load, making it necessary to compute the required section modulus and select the joist cross section accordingly.

Example 1. Determine the joist size required at a spacing of 16 in. to carry a live load of 40 psf on a span of 14 ft. The floor and ceiling construction is as shown in Fig. 10-1. The wood specified for the joists is Douglas fir–larch, No. 2 grade.

Solution: (1) Using data from Table 20-1, we determine the design dead load as follows:

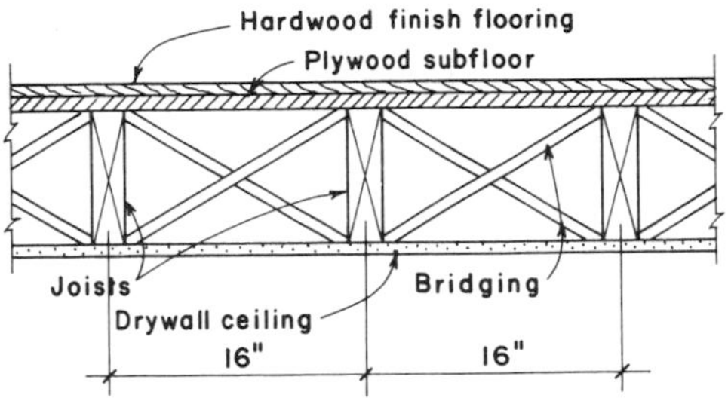

FIGURE 10-1. Typical joist floor construction.

Wood flooring	2.5 psf
$\frac{3}{4}$ in. plywood decking	2.25
Joists and bridging (estimate)	2.75
$\frac{1}{2}$ in. drywall ceiling	2.5
Total dead load	10.0 psf

The total floor loading is thus 40 + 10 = 50 psf. With joists at 16 in. centers, the uniformly distributed load on a single joist is (50)(16/12) = 66.7 lb/lin ft of joist. For the simple span joists the maximum moment is

$$M = \frac{wL^2}{8} = \frac{(66.7)(14)^2}{8} = 1634 \text{ ft-lb}$$

From Table 3-1, the allowable bending stress for the joist is 1450 psi; thus the required section modulus is

$$S = \frac{M}{F_b} = \frac{(1634)(12)}{1450} = 13.52 \text{ in.}^3$$

Inspection of Table 4-1 shows that a 2 × 8 is just short of this requirement and that it is necessary to use a 2 × 10 with S = 21.391 in.^{3}. Alternatives would be to consider using a higher stress grade wood or to place the joists on 12 in. centers.

(2) Shear stress is seldom a critical factor for relatively lightly loaded joists, but an investigation should be made to confirm this.

The maximum shear force is one-half the total load; thus $V = wL/2 = (66.7)(14/2) = 467$ lb. The maximum shear stress on the 2×10 is thus

$$f_v = \frac{3}{2}\frac{V}{bd} = \frac{3(467)}{2(13.875)} = 50.5 \text{ psi}$$

As this is less than the value of F_v, given as 95 psi in Table 3-1, the 2×10 is sufficient.

(3) For this situation the usual deflection limit is for a maximum deflection under live load of 1/360 of the span. We thus determine the following:

$$\text{Allowable deflection} = \frac{L}{360} = \frac{(14 \times 12)}{360} = 0.467 \text{ in.}$$

$$\text{Total live load} = (40)\left(\frac{16}{12}\right)(14) = 747 \text{ lb}$$

$$\text{Maximum } D = \frac{5WL^3}{384EI} = \frac{(5)(747)(14 \times 12)^3}{(384)(1{,}700{,}000)(98.9)} = 0.27 \text{ in.}$$

Since the computed deflection is less than the allowable, the 2×10 is adequate.

Example 2. Floor joists similar to those in Example 1 must carry, in addition to the floor loading, a partition wall that is perpendicular to the joists and 4 ft from one end of the joist span. Investigate the joists designed in Example 1 to see if they are adequate for this additional loading. The partition has a total weight of 120 lb/ft of its length.

Solution: (1) The load on an individual joist is found as $(16/12)(120) = 160$ lb. Using data from Example 1, the loading for a joist is thus as shown in Fig. 10-2, with values of shear and moment as shown on the diagrams.

(2) The maximum bending stress in the joist is

$$f_b = \frac{M}{S} = \frac{(1970 \times 12)}{21.391} = 1105 \text{ psi}$$

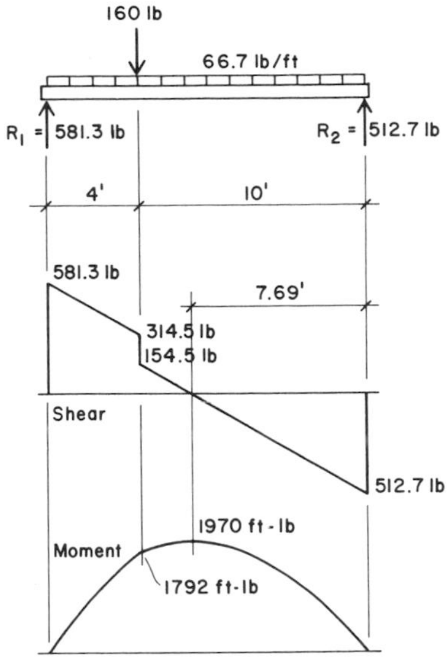

FIGURE 10-2.

As this is less than the maximum allowable stress of 1450 psi, the joist is adequate.

(3) The maximum shear stress in the joist is

$$f_v = \frac{3}{2}\frac{V}{bd} = \frac{(3)(581.3)}{(2)(13.875)} = 62.8 \text{ psi}$$

As this is less than the limit of 95 psi, the joist is adequate.

(4) Live load deflection will be the same as in Example 1 where it was shown to be not critical. If the total deflection under dead and live load is a concern, a reasonably approximate value can be obtained by the use of an equivalent uniform load that will produce the same maximum moment as that under the actual combined loading shown in Fig. 10-2. For this we determine that

$$M = \frac{WL}{8} \quad \text{or} \quad W = \frac{8M}{L}$$

and using the value from Fig. 10-2,

$$W = \frac{8M}{L} = \frac{8(1970)}{14} = 1126 \text{ lb}$$

This load may now be used with the usual deflection formula for a uniformly loaded beam. (See Example 2 in Sec. 8-3.)

10-3 Span Tables for Joists

Many references contain tables from which joists may be selected for ordinary situations. Table 10-1 is a reproduction of Table 25-U-J-1 from the 1985 edition of the *Uniform Building Code,* which provides allowable spans for floor joists for the data given in Example 1 in the preceding section. Referring to that example, the table may be used as follows:

1. The values of $F_b = 1450$ psi and $E = 1{,}700{,}000$ psi are noted for the wood to be used.
2. Using the value for E, enter the table in the vertical column headed by 1.7 (for $E = 1{,}700{,}000$ psi). Note the following possibilities for a span of 14 ft:
   2 × 8 joists at 12 in. centers—14 ft-5 in. span.
   2 × 10 joists at 24 in. centers—14 ft-7 in. span.
   Thus, if the desired spacing of 16 in. is retained, the choice must be for the 2 × 10.
3. If the 16 in. spacing is accepted, note in the boxes in the horizontal row with 2 × 10 at 16 in. that the box showing a span of 14 ft is in the vertical column headed by 1.0 (indicating a required minimum E of 1,000,000 psi).
4. Also note that the box just indicated shows a value for the minimum required allowable bending stress of 920 psi.

If the 2 × 10 joists are used at 16 in. spacing, the wood grade selected is more than adequate, as is verified by the computations in the preceding section.

Span tables for joists are available from many references, including Refs. 2, 3, and 4, in the References following Chapter 20. Other sources include *Architectural Graphic Standards* by Ramsey and Sleeper, Wiley, 1981, and *Design Values for Joists and*

TABLE 10-1. Allowable Spans in Feet and Inches for Floor Joists[a]

Joist Size (in)	Spacing (in)	Modulus of Elasticity, E, in 1,000,000 psi													
		0.8	0.9	1.0	1.1	1.2	1.3	1.4	1.5	1.6	1.7	1.8	1.9	2.0	2.2
2x6	12.0	8-6 720	8-10 780	9-2 830	9-6 890	9-9 940	10-0 990	10-3 1040	10-6 1090	10-9 1140	10-11 1190	11-2 1230	11-4 1280	11-7 1320	11-11 1410
	16.0	7-9 790	8-0 860	8-4 920	8-7 980	8-10 1040	9-1 1090	9-4 1150	9-6 1200	9-9 1250	9-11 1310	10-2 1360	10-4 1410	10-6 1460	10-10 1550
	24.0	6-9 900	7-0 980	7-3 1050	7-6 1120	7-9 1190	7-11 1250	8-2 1310	8-4 1380	8-6 1440	8-8 1500	8-10 1550	9-0 1610	9-2 1670	9-6 1780
2x8	12.0	11-3 720	11-8 780	12-1 830	12-6 890	12-10 940	13-2 990	13-6 1040	13-10 1090	14-2 1140	14-5 1190	14-8 1230	15-0 1280	15-3 1320	15-9 1410
	16.0	10-2 790	10-7 850	11-0 920	11-4 980	11-8 1040	12-0 1090	12-3 1150	12-7 1200	12-10 1250	13-1 1310	13-4 1360	13-7 1410	13-10 1460	14-3 1550
	24.0	8-11 900	9-3 980	9-7 1050	9-11 1120	10-2 1190	10-6 1250	10-9 1310	11-0 1380	11-3 1440	11-5 1500	11-8 1550	11-11 1610	12-1 1670	12-6 1780
2x10	12.0	14-4 720	14-11 780	15-5 830	15-11 890	16-5 940	16-10 990	17-3 1040	17-8 1090	18-0 1140	18-5 1190	18-9 1230	19-1 1280	19-5 1320	20-1 1410
	16.0	13-0 790	13-6 850	14-0 920	14-6 980	14-11 1040	15-3 1090	15-8 1150	16-0 1200	16-5 1250	16-9 1310	17-0 1360	17-4 1410	17-8 1460	18-3 1550
	24.0	11-4 900	11-10 980	12-3 1050	12-8 1120	13-0 1190	13-4 1250	13-8 1310	14-0 1380	14-4 1440	14-7 1500	14-11 1550	15-2 1610	15-5 1670	15-11 1780
2x12	12.0	17-5 720	18-1 780	18-9 830	19-4 890	19-11 940	20-6 990	21-0 1040	21-6 1090	21-11 1140	22-5 1190	22-10 1230	23-3 1280	23-7 1320	24-5 1410
	16.0	15-10 790	16-5 860	17-0 920	17-7 980	18-1 1040	18-7 1090	19-1 1150	19-6 1200	19-11 1250	20-4 1310	20-9 1360	21-1 1410	21-6 1460	22-2 1550
	24.0	13-10 900	14-4 980	14-11 1050	15-4 1120	15-10 1190	16-3 1250	16-8 1310	17-0 1380	17-5 1440	17-9 1500	18-1 1550	18-5 1610	18-9 1670	19-4 1780

[a] *Criteria:* 40 psf live load, 10 psf dead load, live load deflection limited to $\frac{1}{360}$ of the span. Number indicated in each box below the span is the required allowable bending stress in psi.

Source: Reproduced from the *Uniform Building Code,* 1985 edition, (Ref. 3), with permission of the publishers, International Conference of Building Officials.

Rafters, National Forest Products Association. As with any tabular design data, care must be taken to note the criteria on which the table data are based, particularly in this case, the live load, assumed dead load, and deflection limitation.

Problem 10-3-A*. Floor joists of Southern pine, No. 2 grade, are required to carry a live load of 40 psf on a span of 16 ft. Total dead load is approximately 10 psf and live load deflection is limited to $\frac{1}{360}$ of the span. Find (1) the size and required spacing for the shallowest (least depth) joists; (2) the size of joist required if the spacing is 24 in. (Use only spacings of 12, 16, and 24 in.) Use Table 10-1 for this problem.

Problem 10-3-B. Perform computations as in Example 1 of Sec. 10-2 to confirm the choices in Problem 10-3-A.

Problem 10-3-C. For the joists selected in part (2) of Problem 10-3-A, investigate the adequacy of the joists if they support a partition at midspan that is perpendicular to the joists and weighs 140 lb/ft of wall length in addition to the floor loading given in Problem 10-3-A.

10-4 Ceiling Joists

Ceilings are constructed in a number of ways; three common ones are constructions of lath and plaster, of gypsum plasterboard (drywall), and of modular panels attached to a suspended framework. Support for a ceiling may be by direct attachment to the structure above (as shown in Fig. 10-1), by suspension from the structure above, or by an independent set of ceiling joists. If a large attic space is created between a ceiling and the structure above it, it may be necessary to design the ceiling supports as floor joists and to anticipate some storage usage of the attic space. When space is limited to only a few feet, codes generally require only a minimum live load. Table 20-3, from the *Uniform Building Code,* requires a live load of only 10 psf.

For plastered ceilings, the depth of ceiling joists should be conservatively selected to minimize deflection. Although sag is always visually objectionable in an overhead spanning structure, the cracking of the plaster is a critical concern because replacement is quite costly.

Table 10-2 is a reproduction of Table 25-U-J-6 from the *Uniform Building Code* (Ref. 3), and yields spans for ceiling joists

TABLE 10-2. Allowable Spans in Feet and Inches for Ceiling Joists (Drywall Ceiling)[a]

JOIST SIZE (IN)	SPACING (IN)	Modulus of Elasticity, E, in 1,000,000 psi													
		0.8	0.9	1.0	1.1	1.2	1.3	1.4	1.5	1.6	1.7	1.8	1.9	2.0	2.2
2x4	12.0	9-10 710	10-3 770	10-7 830	10-11 880	11-3 930	11-7 980	11-10 1030	12-2 1080	12-5 1130	12.8 1180	12-11 1220	13-2 1270	13-4 1310	13-9 1400
	16.0	8-11 780	9-4 850	9-8 910	9-11 970	10-3 1030	10-6 1080	10-9 1140	11-0 1190	11-3 1240	11-6 1290	11-9 1340	11-11 1390	12-2 1440	12-6 1540
	24.0	7-10 900	8-1 970	8-5 1040	8-8 1110	8-11 1170	9-2 1240	9-5 1300	9-8 1360	9-10 1420	10-0 1480	10-3 1540	10-5 1600	10-7 1650	10-11 1760
2x6	12.0	15-6 710	16-1 770	16-8 830	17-2 880	17-8 930	18-2 980	18-8 1030	19-1 1080	19-6 1130	19-11 1180	20-3 1220	20-8 1270	21-0 1310	21-8 1400
	16.0	14-1 780	14-7 850	15-2 910	15-7 970	16-1 1030	16-6 1080	16-11 1140	17-4 1190	17-8 1240	18-1 1290	18-5 1340	18-9 1390	19-1 1440	19-8 1540
	24.0	12-3 900	12-9 970	13-3 1040	13-8 1110	14-1 1170	14-5 1240	14-9 1300	15-2 1360	15-6 1420	15-9 1480	16-1 1540	16-4 1600	16-8 1650	17-2 1760
2x8	12.0	20-5 710	21-2 770	21-11 830	22-8 880	23-4 930	24-0 980	24-7 1030	25-2 1080	25-8 1130	26-2 1180	26-9 1220	27-2 1270	27-8 1310	28-7 1400
	16.0	18-6 780	19-3 850	19-11 910	20-7 970	21-2 1030	21-9 1080	22-4 1140	22-10 1190	23-4 1240	23-10 1290	24-3 1340	24-8 1390	25-2 1440	25-11 1540
	24.0	16-2 900	16-10 970	17-5 1040	18-0 1110	18-6 1170	19-0 1240	19-6 1300	19-11 1360	20-5 1420	20-10 1480	21-2 1540	21-7 1600	21-11 1650	22-8 1760
2x10	12.0	26-0 710	27-1 770	28-0 830	28-11 880	29-9 930	30-7 980	31-4 1030	32-1 1080	32-9 1130	33-5 1180	34-1 1220	34-8 1270	35-4 1310	36-5 1400
	16.0	23-8 780	24-7 850	25-5 910	26-3 970	27-1 1030	27-9 1080	28-6 1140	29-2 1190	29-9 1240	30-5 1290	31-0 1340	31-6 1390	32-1 1440	33-1 1540
	24.0	20-8 900	21-6 970	22-3 1040	22-11 1110	23-8 1170	24-3 1240	24-10 1300	25-5 1360	26-0 1420	26-6 1480	27-1 1540	27-6 1600	28-0 1650	28-11 1760

[a] *Criteria:* 10 psf live load, 5 psf dead load, live load deflection limited to $\frac{1}{240}$ of the span. Number shown in each box below the span is the minimum required allowable bending stress in psi.

Source: Reproduced from the *Uniform Building Code*, 1985 edition, (Ref. 3), with permission of the publishers, International Conference of Building Officials.

with construction of drywall and a live load of 10 psf. Organization of the table is similar to that for Table 10-1, and the procedure for its use is the same as that described in Sec. 10-3.

Problem 10-4-A,* B, C. Using the following data, select ceiling joists from Table 10-2:

	Joist Spacing	Wood	Span
A	16 in.	Douglas fir–larch, No. 2	12 ft
B	16 in.	Southern pine, No. 3 Dense	14 ft
C	24 in.	Southern pine, No. 2	18 ft

10-5 Design of Rafters

Rafters are the comparatively small, closely spaced beams that support the load on sloping roofs. The span of a rafter is measured along its horizontal projection, as indicated in Fig. 10-3. Although the dimension lines in the figure indicate the distance between centers of supports, it is common practice to consider the span as the clear distance between supports when designing rafters spaced not more than 24 in. on center.

The dead load to be supported consists of the weight of the rafters, the sheathing (roof deck), and the roof covering, such as shingles, tile, or roll roofing. The live load on pitched roofs is the load that results from wind and snow.

The snow load is vertical and the amount of accumulation depends on the roof slope; snow tends to slide off steep roofs but may accumulate on relatively flat ones. Wind action, however,

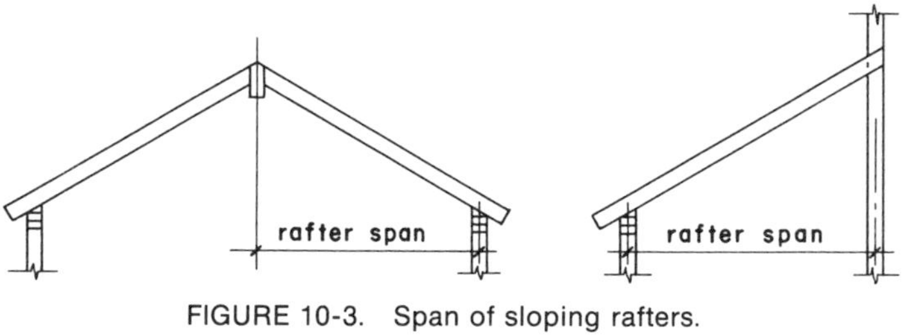

FIGURE 10-3. Span of sloping rafters.

TABLE 10-3. Allowable Spans in Feet and Inches for Low- or High-Slope Rafters (Drywall Ceiling)[a]

Rafter Size (in)	Spacing (in)	Allowable Extreme Fiber Stress in Bending F_b (psi).														
		500	600	700	800	900	1000	1100	1200	1300	1400	1500	1600	1700	1800	1900
2x6	12.0	8-6 0.26	9-4 0.35	10-0 0.44	10-9 0.54	11-5 0.64	12-0 0.75	12-7 0.86	13-2 0.98	13-8 1.11	14-2 1.24	14-8 1.37	15-2 1.51	15-8 1.66	16-1 1.81	16-7 1.96
	16.0	7-4 0.23	8-1 0.30	8-8 0.38	9-4 0.46	9-10 0.55	10-5 0.65	10-11 0.75	11-5 0.85	11-10 0.97	12-4 1.07	12-9 1.19	13-2 1.31	13-7 1.44	13-11 1.56	14-4 1.70
	24.0	6-0 0.19	6-7 0.25	7-1 0.31	7-7 0.38	8-1 0.45	8-6 0.53	8-11 0.61	9-4 0.70	9-8 0.78	10-0 0.88	10-5 0.97	10-9 1.07	11-1 1.17	11-5 1.28	11-8 1.39
2x8	12.0	11-2 0.26	12-3 0.35	13-3 0.44	14-2 0.54	15-0 0.64	15-10 0.75	16-7 0.86	17-4 0.98	18-0 1.11	18-9 1.24	19-5 1.37	20-0 1.51	20-8 1.66	21-3 1.81	21-10 1.96
	16.0	9-8 0.23	10-7 0.30	11-6 0.38	12-3 0.46	13-0 0.55	13-8 0.65	14-4 0.75	15-0 0.85	15-7 0.96	16-3 1.07	16-9 1.19	17-4 1.31	17-10 1.44	18-5 1.56	18-11 1.70
	24.0	7-11 0.19	8-8 0.25	9-4 0.31	10-0 0.38	10-7 0.45	11-2 0.53	11-9 0.61	12-3 0.70	12-9 0.78	13-3 0.88	13-8 0.97	14-2 1.07	14-7 1.17	15-0 1.28	15-5 1.39
2x10	12.0	14-3 0.26	15-8 0.35	16-11 0.44	18-1 0.54	19-2 0.64	20-2 0.75	21-2 0.86	22-1 0.98	23-0 1.11	23-11 1.24	24-9 1.37	25-6 1.51	26-4 1.66	27-1 1.81	27-10 1.96
	16.0	12-4 0.23	13-6 0.30	14-8 0.38	15-8 0.46	16-7 0.55	17-6 0.65	18-4 0.75	19-2 0.85	19-11 0.96	20-8 1.07	21-5 1.19	22-1 1.31	22-10 1.44	23-5 1.56	24-1 1.70
	24.0	10-1 0.19	11-1 0.25	11-11 0.31	12-9 0.38	13-6 0.45	14-3 0.53	15-0 0.61	15-8 0.70	16-3 0.78	16-11 0.88	17-6 0.97	18-1 1.07	18-7 1.17	19-2 1.28	19-8 1.39
2x12	12.0	17-4 0.26	19-0 0.35	20-6 0.44	21-11 0.54	23-3 0.64	24-7 0.75	25-9 0.86	26-11 0.98	28-0 1.11	29-1 1.24	30-1 1.37	31-1 1.51	32-0 1.66	32-11 1.81	33-10 1.96
	16.0	15-0 0.23	16-6 0.30	17-9 0.38	19-0 0.46	20-2 0.55	21-3 0.65	22-4 0.75	23-3 0.85	24-3 0.97	25-2 1.07	26-0 1.19	26-11 1.31	27-9 1.44	28-6 1.56	29-4 1.70
	24.0	12-3 0.19	13-5 0.25	14-6 0.31	15-6 0.38	16-6 0.45	17-4 0.53	18-2 0.61	19-0 0.70	19-10 0.78	20-6 0.88	21-3 0.97	21-11 1.07	22-8 1.17	23-3 1.28	23-11 1.39

[a] *Criteria:* 20 psf live load, 15 psf dead load, live load deflection limited to $\frac{1}{240}$ of the span. Number shown in each box below the span is the minimum required modulus of elasticity in units of 1,000,000 psi. Spans are measured along the horizontal projection and loads are considered as applied along the horizontal projection.

Source: Reproduced from the *Uniform Building Code*, 1985 edition, (Ref. 3), with permission of the publishers, International Conference of Building Officials.

TABLE 10-4. Allowable Spans in Feet and Inches for Low-Slope Rafters, Slope 3 in 12 or Less (No Ceiling)[a]

Rafter Size (in)	Spacing (in)	Allowable Extreme Fiber Stress in Bending F_b (psi).														
		500	600	700	800	900	1000	1100	1200	1300	1400	1500	1600	1700	1800	1900
2x6	12.0	9-2 0.33	10-0 0.44	10-10 0.55	11-7 0.67	12-4 0.80	13-0 0.94	13-7 1.09	14-2 1.24	14-9 1.40	15-4 1.56	15-11 1.73	16-5 1.91	16-11 2.09	17-5 2.28	17-10 2.47
	16.0	7-11 0.29	8-8 0.38	9-5 0.48	10-0 0.58	10-8 0.70	11-3 0.82	11-9 0.94	12-4 1.07	12-10 1.21	13-3 1.35	13-9 1.50	14-2 1.65	14-8 1.81	15-1 1.97	15-6 2.14
	24.0	6-6 0.24	7-1 0.31	7-8 0.39	8-2 0.48	8-8 0.57	9-2 0.67	9-7 0.77	10-0 0.88	10-5 0.99	10-10 1.10	11-3 1.22	11-7 1.35	11-11 1.48	12-4 1.61	12-8 1.75
2x8	12.0	12-1 0.33	13-3 0.44	14-4 0.55	15-3 0.67	16-3 0.80	17-1 0.94	17-11 1.09	18-9 1.24	19-6 1.40	20-3 1.56	20-11 1.73	21-7 1.91	22-3 2.09	22-11 2.28	23-7 2.47
	16.0	10-6 0.29	11-6 0.38	12-5 0.48	13-3 0.58	14-0 0.70	14-10 0.82	15-6 0.94	16-3 1.07	16-10 1.21	17-6 1.35	18-2 1.50	18-9 1.65	19-4 1.81	19-10 1.97	20-5 2.14
	24.0	8-7 0.24	9-4 0.31	10-1 0.39	10-10 0.48	11-6 0.57	12-1 0.67	12-8 0.77	13-3 0.88	13-9 0.99	14-4 1.10	14-10 1.22	15-3 1.35	15-9 1.48	16-3 1.61	16-8 1.75
2x10	12.0	15-5 0.33	16-11 0.44	18-3 0.55	19-6 0.67	20-8 0.80	21-10 0.94	22-10 1.09	23-11 1.24	24-10 1.40	25-10 1.56	26-8 1.73	27-7 1.91	28-5 2.09	29-3 2.28	30-1 2.47
	16.0	13-4 0.29	14-8 0.38	15-10 0.48	16-11 0.58	17-11 0.70	18-11 0.82	19-10 0.94	20-8 1.07	21-6 1.21	22-4 1.35	23-2 1.50	23-11 1.65	24-7 1.81	25-4 1.97	26-0 2.14
	24.0	10-11 0.24	11-11 0.31	12-11 0.39	13-9 0.48	14-8 0.57	15-5 0.67	16-2 0.77	16-11 0.88	17-7 0.99	18-3 1.10	18-11 1.22	19-6 1.35	20-1 1.48	20-8 1.61	21-3 1.75
2x12	12.0	18-9 0.33	20-6 0.44	22-2 0.55	23-9 0.67	25-2 0.80	26-6 0.94	27-10 1.09	29-1 1.24	30-3 1.40	31-4 1.56	32-6 1.73	33-6 1.91	34-7 2.09	35-7 2.28	36-7 2.47
	16.0	16-3 0.29	17-9 0.38	19-3 0.48	20-6 0.58	21-9 0.70	23-0 0.82	24-1 0.94	25-2 1.07	26-2 1.21	27-2 1.35	28-2 1.50	29-1 1.65	29-11 1.81	30-10 1.97	31-8 2.14
	24.0	13-3 0.24	14-6 0.31	15-8 0.39	16-9 0.48	17-9 0.57	18-9 0.67	19-8 0.77	20-6 0.88	21-5 0.99	22-2 1.10	23-0 1.22	23-9 1.35	24-5 1.48	25-2 1.61	25-10 1.75

[a] *Criteria:* 20 psf live load, 10 psf dead load, live load deflection limited to $\frac{1}{240}$ of the span. Number shown in each box below the span is the minimum required modulus of elasticity in units of 1,000,000 psi. Spans are measured along the horizontal projection and loads are considered as applied along the horizontal projection.

Source: Reproduced from the *Uniform Building Code*, 1985 edition, (Ref. 3), with permission of the publishers, International Conference of Building Officials.

TABLE 10-5. Allowable Spans in Feet and Inches for High-Slope Rafters, Slope Over 3 in 12 (Light Roof Covering)[a]

Rafter Size (in)	Spacing (in)	Allowable Extreme Fiber Stress in Bending F_b (psi).														
		500	600	700	800	900	1000	1100	1200	1300	1400	1500	1600	1700	1800	1900
2x4	12.0	6-2 0.29	6-9 0.38	7-3 0.49	7-9 0.59	8-3 0.71	8-8 0.83	9-1 0.96	9-6 1.09	9-11 1.23	10-3 1.37	10-8 1.52	11-0 1.68	11-4 1.84	11-8 2.00	12-0 2.17
	16.0	5-4 0.25	5-10 0.33	6-4 0.42	6-9 0.51	7-2 0.61	7-6 0.72	7-11 0.83	8-3 0.94	8-7 1.06	8-11 1.19	9-3 1.32	9-6 1.45	9-10 1.59	10-1 1.73	10-5 1.88
	24.0	4-4 0.21	4-9 0.27	5-2 0.34	5-6 0.42	5-10 0.50	6-2 0.59	6-5 0.68	6-9 0.77	7-0 0.87	7-3 0.97	7-6 1.08	7-9 1.19	8-0 1.30	8-3 1.41	8-6 1.53
2x6	12.0	9-8 0.29	10-7 0.38	11-5 0.49	12-3 0.59	13-0 0.71	13-8 0.83	14-4 0.96	15-0 1.09	15-7 1.23	16-2 1.37	16-9 1.52	17-3 1.68	17-10 1.84	18-4 2.00	18-10 2.17
	16.0	8-4 0.25	9-2 0.33	9-11 0.42	10-7 0.51	11-3 0.61	11-10 0.72	12-5 0.83	13-0 0.94	13-6 1.06	14-0 1.19	14-6 1.32	15-0 1.45	15-5 1.59	15-11 1.73	16-4 1.88
	24.0	6-10 0.21	7-6 0.27	8-1 0.34	8-8 0.42	9-2 0.50	9-8 0.59	10-2 0.68	10-7 0.77	11-0 0.87	11-5 0.97	11-10 1.08	12-3 1.19	12-7 1.30	13-0 1.41	13-4 1.53
2x8	12.0	12-9 0.29	13-11 0.38	15-1 0.49	16-1 0.59	17-1 0.71	18-0 0.83	18-11 0.96	19-9 1.09	20-6 1.23	21-4 1.37	22-1 1.52	22-9 1.68	23-6 1.84	24-2 2.00	24-10 2.17
	16.0	11-0 0.25	12-1 0.33	13-1 0.42	13-11 0.51	14-10 0.61	15-7 0.72	16-4 0.83	17-1 0.94	17-9 1.06	18-5 1.19	19-1 1.32	19-9 1.45	20-4 1.59	20-11 1.73	21-6 1.88
	24.0	9-0 0.21	9-10 0.27	10-8 0.34	11-5 0.42	12-1 0.50	12-9 0.59	13-4 0.68	13-11 0.77	14-6 0.87	15-1 0.97	15-7 1.08	16-1 1.19	16-7 1.30	17-1 1.41	17-7 1.53
2x10	12.0	16-3 0.29	17-10 0.38	19-3 0.49	20-7 0.59	21-10 0.71	23-0 0.83	24-1 0.96	25-2 1.09	26-2 1.23	27-2 1.37	28-2 1.52	29-1 1.68	30-0 1.84	30-10 2.00	31-8 2.17
	16.0	14-1 0.25	15-5 0.33	16-8 0.42	17-10 0.51	18-11 0.61	19-11 0.72	20-10 0.83	21-10 0.94	22-8 1.06	23-7 1.19	24-5 1.32	25-2 1.45	25-11 1.59	26-8 1.73	27-5 1.88
	24.0	11-6 0.21	12-7 0.27	13-7 0.34	14-6 0.42	15-5 0.50	16-3 0.59	17-1 0.68	17-10 0.77	18-6 0.87	19-3 0.97	19-11 1.08	20-7 1.19	21-2 1.30	21-10 1.41	22-5 1.53

[a] *Criteria:* 20 psf live load, 7 psf dead load, live load deflection limited to $\frac{1}{180}$ of the span. Number in each box below the span is the minimum required modulus of elasticity in units of 1,000,000 psi. Spans are measured along the horizontal projection and loads are considered as applied along the horizontal projection.

Source: Reproduced from the *Uniform Building Code,* 1985 edition, (Ref. 3), with permission of the publishers, International Conference of Building Officials.

produces loads with a horizontal component; and because it is customary to consider that the wind acts in a direction perpendicular to the roof surface the magnitude of the wind pressure becomes smaller as the slope of the roof becomes flatter. It is generally recognized as unreasonable to expect that full snow and wind loads will occur simultaneously on a roof; if the wind were blowing hard enough to produce the maximum pressure for which the roof is designed, much of the snow would be blown off. This and other considerations have led to the use of an *equivalent vertical load* for the combined snow and wind loads on rafter roofs. Some building codes specify that the equivalent vertical load be considered uniformly distributed over the actual area of roof surface, whereas others establish values that permit the horizontal projection of the area (as in rafter spans) to be used.

Design of relatively short span rafters is mostly done with load-span tables. Besides the usual variables, the angle of slope of the rafters must be considered. Tables 10-3, 10-4, and 10-5 present design data for three common situations, and are reproduced from a series presented in the *Uniform Building Code* (Ref. 3). For various reasons, these tables are organized somewhat differently from the joist tables shown in the preceding sections. In this case the vertical columns in the tables are headed by the allowable bending stress, and the corresponding minimum required modulus of elasticity is given in each box. The procedure for their use, however, is essentially the same as that described in Sec. 10-3.

In many instances it is possible to use an increase in allowable bending stress based on load duration, as described in Sec. 3-3. If the building code permits this for the situation, the modification should be made on the value obtained from Table 3-1 and the modified value should then be used in the span tables. The modulus of elasticity, however, is not modified in these cases.

Care should be taken to use the table with appropriate criteria, including those for roof slope, live load, and ceiling construction. As discussed for the joist span tables, extensive tabulations are available in various references, including Refs. 2, 3, and 4 in the References of this book.

Problem 10-5-A*. Using Table 10-3, select the minimum size rafter for a span of 18 ft if rafters are 16 in. on center and the wood is Douglas fir–larch, No. 2 grade.

Problem 10-5-B. Using Table 10-4, select the minimum size rafter for a span of 20 ft if rafters are 16 in. on center and the wood is Douglas fir–larch, No. 1 grade.

Problem 10-5-C. Using Table 10-5, select the minimum size rafter for a span of 24 ft if rafters are 16 in. on center and the wood grade is Douglas fir–larch, No. 1 grade.

11

Board and Plank Decks

11-1 Board Decks

Before plywood established its dominance as a decking material, most roof and floor decks were made with $\frac{3}{4}$ in. (nominal 1 in.) boards, usually with interlocking edges created by tongue–and–groove joints, as shown in Fig. 11-1*a*. Today this type of deck is used only in regions where labor cost is relatively low and the boards are locally competitive in availability and cost in comparison to structural plywood.

When installed in a position with the boards perpendicular to the supports, board decks produce a rather poor horizontal diaphragm. It is common, therefore, when significant diaphragm action for lateral loads is required, to install the deck at an angle of 45° to the joists, resulting in the development of a truss action with the support framing.

Decks of nominal 1 in. boards are usually adequate for roofs and floors where spacing of rafters or joists does not exceed 24 in. However, the type of roofing or the type of finish flooring to be used must be considered. Roofing of all types must be anchored to the deck, usually with nails of some kind. Board decks are usually adequate for holding of nails—possibly better in this re-

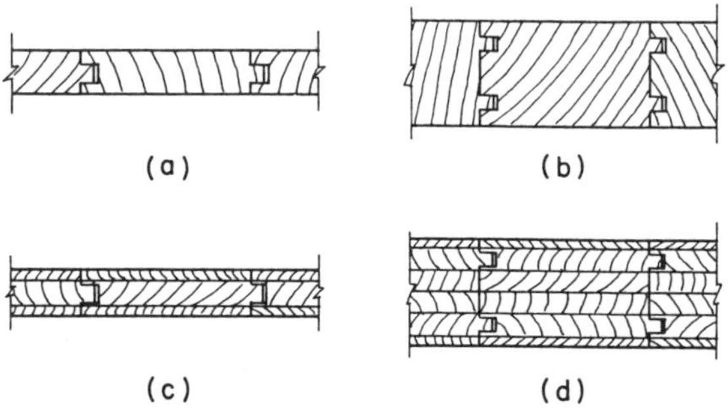

FIGURE 11-1. Units for board and plank decks.

gard than the thinner plywoods. Membrane roofing for flat roofs usually requires a minimum of $\frac{1}{2}$ in. plywood, thus making the board deck more competitive than in the case of sloping roofs with shingles that may be achieved with thinner plywoods.

For floors it is common to use some additional material on top of the structural deck, such as a thin layer of concrete fill or particleboard sheets. These filler materials add considerable stiffness to the deck and so when they are *not* to be used, some conservative judgment should be used in choosing deck thickness and support spacing.

11-2 Plank Decks

If a deck of the board type is thicker than $\frac{3}{4}$ in., it is generally referred to as *planking* or a *plank deck*. The most widely used form of plank deck is that made with 2 in. nominal thickness units (approximately 1.5 in. in actual thickness). There are usually specific reasons for selecting such a deck, including one or all of the following:

1. The deck is to be exposed on the underside and the appearance of the planks is considerably better than that of a plywood deck of typical structural units.
2. Exposed to view or not, the deck may require a fire rating and the thicker decks are much better in this regard. The 2

in. nominal thickness is usually the minimum required with construction qualified as "heavy timber" for fire rating.

3. It may be desired to have supporting members with spacing exceeding that which is feasible for board or plywood decking.
4. Concentrated loadings from vehicles or equipment may be too high for thinner decks.

Nominal 2 in. plank deck may be of the same form as board deck (Fig. 11-1*a*), but is often made with laminated units as shown in Fig. 11-1*c*. Plank deck is also available in the thicknesses greater than 2 in. nominal. When thickness exceeds 2.5 in. or so, the units usually have a double tongue–and–groove on each face, as shown in Fig. 11-1*b* for a solid unit and in Fig. 11-1*d* for a laminated unit. The thicker plank deck units are capable of achieving considerable span distances, and may be used in structures without the usual rafters or joists, such as decks spanning from wall–to–wall or from beam–to–beam in a widely spaced framing system.

One problem with plank decks—as with board decks—is the low diaphragm capacity for resistance of lateral loads when the deck units are perpendicular to supports. Although diagonal placement—as discussed with board decks—is a possibility, it is less common than in the case of the $\frac{3}{4}$ in. decks. When diaphragm capacity of some significant magnitude is required, the usual solution is to nail plywood sheets to the top of the planks. This is quite commonly done, and is frankly the main reason that diaphragm capacities are given for decks of plywood as thin as $\frac{5}{16}$ in.—a thickness not usually usable for a structural spanning deck. (See *Uniform Building Code,* Table 25-J-1, reproduced as Table 19-1 here.)

Four types of spans for plank floors are generally recognized: simple, two-span continuous, combination simple and two-span continuous, and controlled random. The last three types are all stiffer, in varying degree, than the simple span because of the continuity introduced by the different arrangements of the pieces of planking. The four span types are identified in Fig. 11-2. Examination of the figure shows that all planks are the same length in

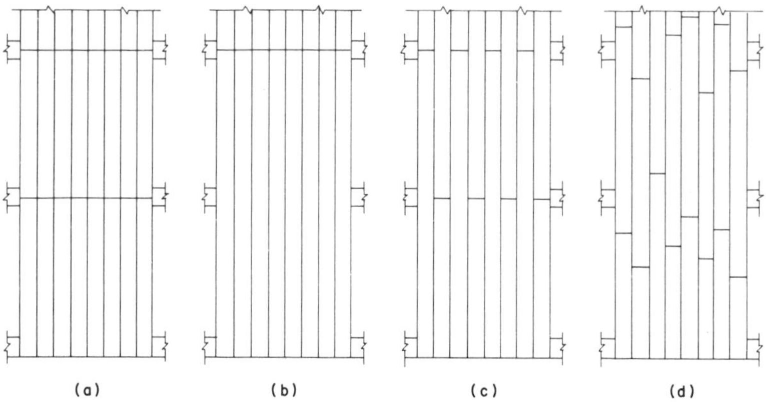

FIGURE 11-2. Plank deck installation and span conditions: (*a*) simple span; (*b*) two-span continuous; (*c*) combination simple and two-span continuous; (*d*) controlled random length.

the simple span with end joints over each beam. The plank lengths are also equal for the two-span continuous arrangement with end joints over every other beam. For the combination span all pieces are two spans in length except for every other piece in the end span, but the end joints over intermediate beams are staggered in adjacent lines of decking. The random arrangement permits the use of economical random lengths of decking. The principal control requirement of this type of span is that end joints be well scattered and each piece of plank bear on at least one beam.

Plank decks are essentially fabricated products and information about them should be obtained from the suppliers or manufacturers of the products. Type of units, finishes of exposed undersides, installation specifications, and structural capacities vary greatly and the products available regionally should be used for design work.

12

Wood Columns

12-1 Introduction

A column is a compression member, the length of which is several times greater than its least lateral dimension. The term *column* is generally applied to relatively heavy vertical members and the term *strut* is given to smaller compression members not necessarily in a vertical position. The type of wood column used most frequently is the *simple solid column,* which consists of a single piece of wood that is square or rectangular in cross section. Solid columns of circular cross section are also considered simple solid columns, but they are used less frequently. A *spaced column* is an assembly of two or more pieces with their longitudinal axes parallel and separated at the ends and middle points of their length by blocking. Two other types are *built-up columns* with mechanical fastenings and *glued-laminated columns*. The *studs* in light wood framing are also columns.

12-2 Slenderness Ratio

In wood construction the slenderness ratio of a freestanding simple solid column is the ratio of the unbraced (laterally unsup-

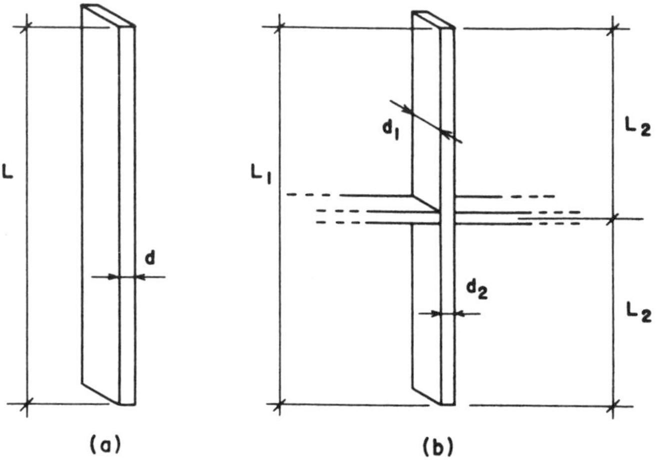

FIGURE 12-1.

ported) length to the dimension of its least side, or L/d. (Fig. 12-1*a*.) When members are braced so that the unsupported length with respect to one face is less than that with respect to the other, L is the distance between the points of support that prevent lateral movement in the direction along which the dimension of the section is measured. This is illustrated in Fig. 12-1*b*. If the section is not square or round, it may be necessary to investigate two L/d conditions for such a column to determine which is the limiting one. The slenderness ratio for simple solid columns is limited to $L/d = 50$; for spaced columns the limiting ratio is $L/d = 80$.

12-3 Capacity of Simple Solid Columns

Figure 12-2 illustrates the typical form of the relationship between axial compression capacity and slenderness for a linear compression member (column). The two limiting conditions are those of the very short member and the very long member. The short member (such as a block of wood) fails in crushing, which is limited by the mass of material and the stress limit in compression. The very long member (such as a yardstick) fails in elastic buckling, which is determined by the stiffness of the member; stiffness is determined by a combination of geometric property

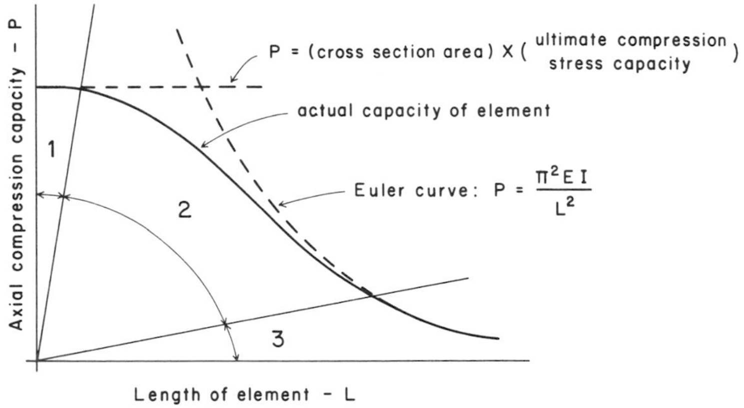

FIGURE 12-2. Relation of column length to axial compression capacity.

(shape of the cross section) and material stiffness property (modulus of elasticity). Between these two extremes—which is where most wood compression members fall—the behavior is indeterminate because the transition is made between the two distinctly different modes of behavior.

The NDS currently provides for three separate compression stress calculations, corresponding to the three zones of behavior described in Fig. 12-2. The plot of these three stress formulas, for a specific example wood, is shown in Fig. 12-3. Typical analysis and design procedures for simple solid wood columns are illustrated in the following examples.

Example 1. A wood compression member consists of a 3 × 6 of Douglas fir–larch, Dense No. 1 grade. Find the allowable axial compression force for unbraced lengths of (1) 2 ft [0.61 m]; (2) 4 ft [1.22 m]; and (3) 8 ft [2.44 m].

Solution: We find from Table 13-1 that $F_c = 1450$ psi [10.0 MPa] and $E = 1{,}900{,}000$ psi [13.1 GPa]. To establish the zone limits we compute the following:

$$11(d) = 11(2.5) = 27.5 \text{ in.}$$

$$50(d) = 50(2.5) = 125 \text{ in.}$$

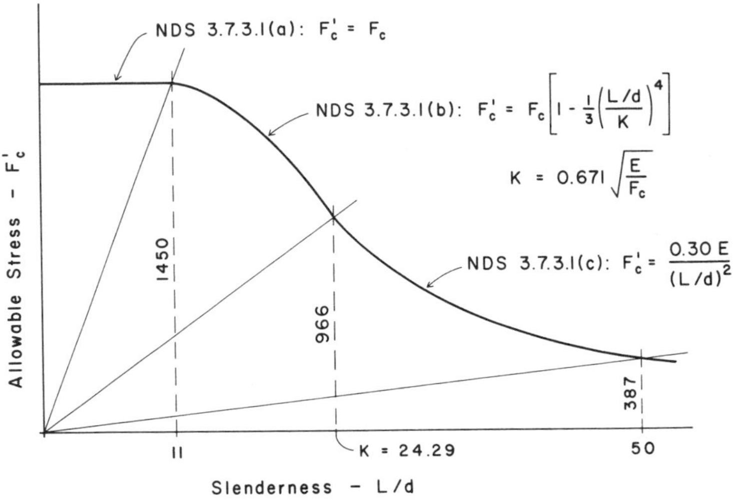

FIGURE 12-3. Allowable axial compression stress as a function of the slenderness ratio *L/d*. National Design Specification (NDS) requirements for Douglas fir–larch, Dense No. 1 grade.

and

$$K = 0.671\sqrt{\frac{E}{F_c}} = 0.671\sqrt{\frac{1{,}900{,}000}{1450}} = 24.29$$

Thus for condition 1, $L = 24$ in., which is in zone 1; $F'_c = F_c = 1450$ psi; allowable $C = F'_c \times$ gross area $= 1450 \times 13.75 = 19{,}938$ lb [88.7 kN].

For condition 2, $L = 48$ in.; $L/d = 48/2.5 = 19.2$, which is in zone 2.

$$F'_c = F_c\left\{1 - \frac{1}{3}\left(\frac{L/d}{K}\right)^4\right\}$$

$$= 1450\left\{1 - \frac{1}{3}\left(\frac{19.2}{24.29}\right)^4\right\}$$

$$= 1262 \text{ psi}$$

Allowable $C = 1262 \times (13.75) = 17{,}353$ lb [77.2 kN].

For condition 3, $L = 96$ in.; $L/d = 96/2.5 = 38.4$, which is in zone 3.

$$F_c' = \frac{0.3(E)}{(L/d)^2} = \frac{0.3(1{,}900{,}000)}{(38.4)^2} = 387 \text{ psi}$$

Allowable compression $= 387 \times (13.75) = 5321$ lb [23.7 kN].

Example 2. Wood 2 × 4 elements are to be used as vertical compression members to form a wall (ordinary stud wall construction). If the wood is Douglas fir–larch, No. 3 grade, and the wall is 8.5 ft high, what is the column load capacity of a single stud?

Solution: In this case it must be assumed that the surfacing materials used for the wall (plywood, dry wall, plaster, etc.) will provide adequate bracing for the studs on their weak axis (the 1.5 in. [38-mm] direction). If not, the studs cannot be used, since the specified height of the wall is considerably in excess of the limit for L/d for a solid column (50). We therefore assume the direction of potential buckling to be that of the 3.5 in. [89-mm] dimension. Thus

$$\frac{L}{d} = \frac{8.5 \times 12}{3.5} = 29.1$$

To determine which column load formula must be used we must find the value of K for this wood. From Table 13-1 we find $F_c = 600$ psi [4.65 MPa] and $E = 1{,}500{,}000$ psi [10.3 GPa]. Then

$$K = 0.671\sqrt{\frac{E}{F_c}} = 0.671\sqrt{\frac{1{,}500{,}000}{600}} = 33.55$$

We thus establish the condition for the stud as zone 2 (Fig. 12-2), and the allowable compression stress is computed as

$$F_c' = F_c\left\{1 - \frac{1}{3}\left(\frac{L/d}{K}\right)^4\right\}$$

$$= 600\left\{1 - \frac{1}{3}\left(\frac{29.1}{33.55}\right)^4\right\} = 467 \text{ psi [3.36 MPa]}$$

The allowable load for the stud is

$$P = F_c' \times \text{gross area} = 487 \times 5.25 = 2557 \text{ lb } [11.4 \text{ kN}]$$

Example 3. A wood column of Douglas fir–larch, Dense No. 1 grade, must carry an axial load of 40 kips [178 kN]. Find the smallest section for unbraced lengths of (1) 4 ft [1.22 m]; (2) 8 ft [2.44 m]; and (3) 16 ft (4.88 m].

Solution: Since the size of the column is unknown, the values of L/d, F_c, and E cannot be predetermined. Therefore, without design aids (tables, graphs, or computer programs), the process becomes a cut-and-try approach in which a specific value is assumed for d and the resulting values for L/d, F_c, E, and F_c' are determined. A required area is then determined and the sections with the assumed d compared with the requirement. If an acceptable member cannot be found, another try must be made with a different d. Although somewhat clumsy, the process is usually not all that laborious, since a limited number of available sizes are involved.

We first consider the possibility of a zone 1 stress condition (Fig. 12-2) since this calculation is quite simple. If the maximum $L = 11(d)$, then the minimum $d = (4 \times 12)/11 = 4.36$ in. [111 mm]. This requires a nominal thickness of 6 in., which puts the size range into the "posts and timbers" category in Table 3-1 for which the allowable stress F_c is 1200 psi. The required area is thus

$$A = \frac{\text{load}}{F_c'} = \frac{40{,}000}{1200} = 33.3 \text{ in.}^2 \; [21{,}485 \text{ mm}^2]$$

The smallest section is thus a 6 × 8, with an area of 41.25 in.^{2}, since a 6 × 6 with 30.25 in.^{2} is not sufficient. (See Table 4-1.) If the oblong-shape column is acceptable, this becomes the smallest member usable. If a square shape is desired, the smallest size would be an 8 × 8.

If the 6 in. nominal thickness is used for the 8 ft column, we determine that

$$\frac{L}{d} = \frac{8 \times 12}{5.5} = 17.45$$

Since this is greater than 11, the allowable stress is in the next zone for which

$$F'_c = F_c\left\{1 - \frac{1}{3}\left(\frac{L/d}{K}\right)^4\right\}$$

$$= 1200\left\{1 - \frac{1}{3}\left(\frac{17.45}{25.26}\right)^4\right\}$$

$$= 1109 \text{ psi } [7.65 \text{ MPa}]$$

in which

$$F_c = 1200 \text{ psi and } E = 1{,}700{,}000 \text{ (from Table 3-1)}$$

and

$$K = 0.671\sqrt{\frac{E}{F_c}} = 0.671\sqrt{\frac{1{,}700{,}000}{1200}} = 25.26$$

The required area is thus

$$A = \frac{\text{load}}{F'_c} = \frac{40{,}000}{1109} = 36.07 \text{ in.}^2\ [23{,}272 \text{ mm}^2]$$

and the choices remain the same as for the 4 ft column.

If the 6 in. nominal thickness is used for the 16 ft column, we determine that

$$\frac{L}{d} = \frac{16 \times 12}{5.5} = 34.9$$

Since this is greater than the value of K, the stress condition is that of zone 3 (Fig. 12-2), and the allowable stress is

$$F'_c = \frac{0.30E}{(L/d)^2} = \frac{(0.30)(1{,}700{,}000)}{34.9^2} = 419 \text{ psi } [2.89 \text{ MPa}]$$

which requires an area for the column of

$$A = \frac{\text{load}}{F'_c} = \frac{40{,}000}{419} = 95.5 \text{ in.}^2 \text{ [61,617 mm}^2\text{]}$$

This is greater than the area for the largest section with a nominal thickness of 6 in. as listed in Table 4-1. Although larger sections may be available in some areas, it is highly questionable to use a member with these proportions as a column. Therefore we consider the next larger nominal thickness of 8 in. Then, if

$$\frac{L}{d} = \frac{16 \times 12}{7.5} = 25.6$$

we are still in the zone 3 condition and the allowable stress is

$$F'_c = \frac{0.30E}{(L/d)^2} = \frac{(0.30)(1{,}700{,}000)}{25.6^2} = 778 \text{ psi [5.36 MPa]}$$

which requires an area of

$$A = \frac{\text{load}}{F'_c} = \frac{40{,}000}{778} = 51.4 \text{ in.}^2 \text{ [33,163 mm}^2\text{]}$$

The smallest member usable is thus an 8 × 8. It is interesting to note that the required square column remains the same for all the column lengths even though the stress varies from 1200 psi to 778 psi. This is not uncommon and is simply due to the limited number of sizes available for the square column section.

(*Note:* For the following problems use Douglas fir–larch, No. 1 grade.)

Problems 12-3-A*-B-C-D. Find the allowable axial compression load for each of the following:

	Nominal Size (in.)	Unbraced Length (ft)	Unbraced Length (m)
A	4 × 4	8	2.44
B	6 × 6	12	3.66
C	8 × 8	18	5.49
D	8 × 8	14	4.27

Problems 12-3-E*-F-G-H. Select the smallest square section for each of the following:

	Axial Load		Unbraced Length	
	(kips)	(kN)	(ft)	(m)
E	20	89	8	2.44
F	50	222	12	3.66
G	50	222	20	6.10
H	100	445	16	4.88

12-4 Use of Design Aids for Wood Columns

It should be apparent from these examples that the design of wood columns by these procedures is a laborious task. The working designer therefore typically utilizes some design aids in the form of graphs, tables, or computer-aided processes. One should exercise care in using such aids, however, to be sure that any specific values for E or F_c that are used correspond to the true conditions of the design work and that the aids are developed from criteria identical to those in any applicable code for the work.

Figure 12-4 consists of a graph from which the axial compression capacity of solid, square wood columns may be determined. Note that the graph curves are based on a specific species and grade of wood (Douglas fir–larch, Dense No. 1 grade). The three circled points on the graph correspond to the design examples in Example 3 of Sec. 12-3.

Table 12-1 gives the axial compression capacity for a range of sizes and unbraced lengths of solid, rectangular wood sections. Note that the design values for elements with nominal thickness of 4 in. and less are different from those with nominal thickness of 6 in. and over, owing to the different size classifications as given in Table 3-1.

Problems 12-4-A-B-C-D. Select square column sections for the loading and lateral bracing conditions given for Problems 12-3-E-F-G-H using data from Table 12-1. Note that in the problems in Sec. 12-3, the wood is No. 1 grade, whereas Table 12-1 uses Dense No. 1 grade. It is possible therefore that the selections from the table may not agree with those made from the computations.

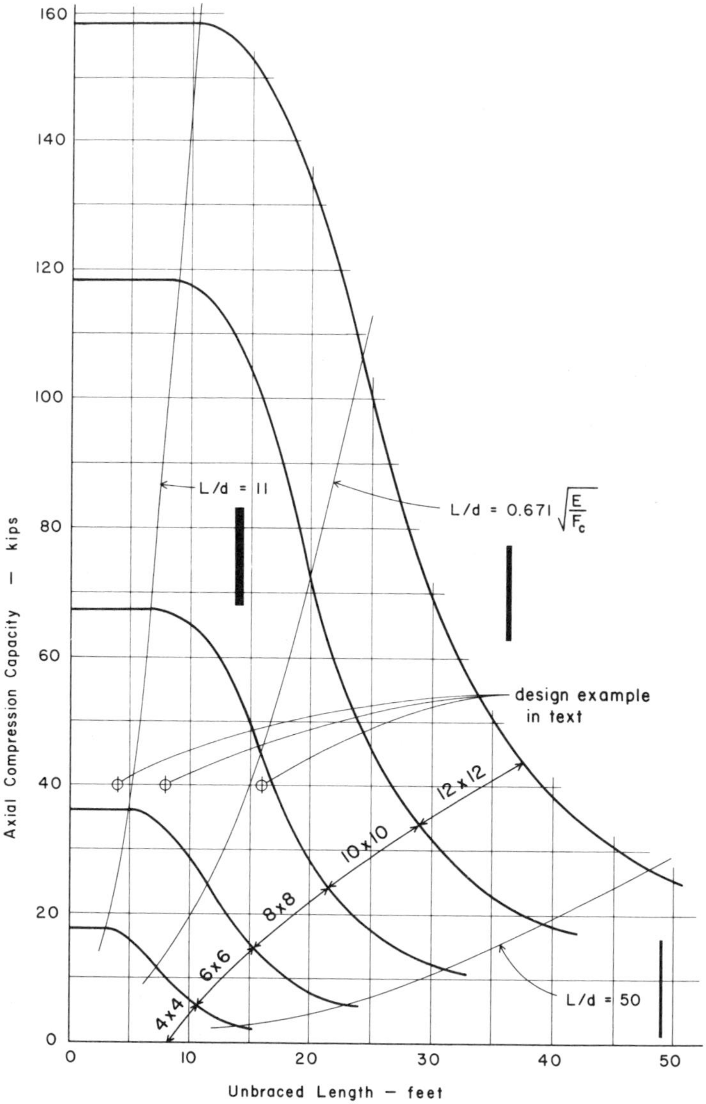

FIGURE 12-4. Axial compression capacity for wood columns with square cross section. Derived from National Design Specification (NDS) requirements for Douglas fir–larch, Dense No. 1 grade.

TABLE 12-1. Capacity of Solid Wood Columns (kips)[a]

Element Size		Unbraced Length (ft)							
Designation	Area of Section (in²)	6	8	10	12	14	16	18	20
2 × 3	3.375	0.8			*L/d* greater than 50				
2 × 4	5.25	1.3							
3 × 4	8.75	6.0	3.4	2.2					
3 × 6	13.75	9.4	5.3	3.4					
4 × 4	12.25	14.7	9.3	5.9	4.1	3.0			
4 × 6	19.25	23.1	14.6	9.3	6.5	4.8			
4 × 8	25.375	30.4	19.2	12.3	8.5	6.3			
6 × 6	30.25	35.4	33.5	29.6	22.5	16.6	12.7	10.0	8.1
6 × 8	41.25	48.3	45.7	40.3	30.6	22.6	17.3	13.6	11.0
6 × 10	52.25	61.2	57.9	51.1	38.8	28.6	21.9	17.2	14.0
6 × 12	63.25	74.1	70.1	61.8	47.0	34.7	26.5	20.9	17.0
8 × 8	56.25	67.5	66.0	63.9	60.0	53.6	43.8	34.6	28.0
8 × 10	71.25	85.5	83.6	80.9	75.9	67.9	55.4	46.9	35.5
8 × 12	86.25	103.5	101.2	98.0	91.9	82.2	67.1	53.0	42.9
8 × 14	101.25	121.5	118.9	115.0	107.9	96.5	78.8	62.3	50.4
10 × 10	90.25	108.3	108.3	106.0	103.6	99.6	93.5	84.5	72.1
10 × 12	109.25	131.1	131.1	128.4	125.4	120.6	113.2	102.4	87.3
10 × 14	128.25	153.9	153.9	150.7	147.2	141.6	132.9	120.2	102.5
10 × 16	147.25	176.7	176.7	173.0	169.0	162.5	152.5	138.0	117.7
12 × 12	132.25	158.7	158.7	158.7	155.5	152.7	148.6	142.5	134.1

Note: Wood used is dense No. 1 Douglas fir–larch under normal moisture and load conditions.

[a] For Douglas fir–larch, Dense No. 1 grade under normal moisture and load duration conditions.

12-5 Round Columns

Solid wood columns of circular cross section are not used extensively in general building construction. As for load-bearing capacity, round and square wood columns of the same cross-sectional area will support the same axial loads and have the same degree of stiffness.

When designing a wood column of circular cross section, a simple procedure is to design a square column first and then select a round column with an equivalent cross-sectional area. To find the diameter of the equivalent round column the dimension *d* of the square column is multiplied by 1.128.

12-6 Poles

Poles are round timbers consisting of the peeled logs of coniferous trees. In short lengths they may be relatively constant in diameter, but when long they are tapered in form, which is the natural form of the tree trunk. As columns, poles are designed with the same basic criteria used for sawn sections. For slenderness considerations the *d* used is taken as that of a square section of equal area. Thus, calling the pole diameter *D*,

$$d^2 = \frac{\pi D^2}{4}, \qquad d = \sqrt{0.7854D^2} = 0.886D$$

For a tapered column, a conservative assumption for design is that the critical column diameter is the least diameter at the small end. If the column is very short, this is reasonable. However, for a slender column, with buckling occurring in the midheight of the column, this is very conservative and the code provides for some adjustment. Nevertheless, because of a typical lack of initial straightness and presence of numerous flaws, many designers prefer to use the unadjusted small end diameter for design computations.

Timber poles are used for foundations in the form of driven piles. They are also used as buried-end posts for fences, signs, and utility transmission lines. Buildings of so called *pole construction* typically use buried-end posts as building columns. For lateral forces, buried-end posts must be designed for bending and—if sustaining significant vertical loads—as combined action members with bending plus axial compression. For the latter situations it is common to consider the round pole to be equivalent to a square sawn section with the same cross section area.

The wood-framed structure utilizing poles has a long history of use and is still a practical solution for utilitarian buildings where good poles are readily available. Accepted practices of construction for these buildings are largely based on experience and do not always yield to rational engineering analysis.

12-7 Stud Wall Construction

Studs are the vertical elements used for wall framing in light wood construction. Studs serve utilitarian purposes of providing for attachment of wall surfacings, but may also serve as vertical columns when the wall provides bearing support for roof or floor systems. The most common stud is the 2 × 4 spaced at intervals of 12, 16, or 24 in.

Studs of nominal 2 in. thickness must be braced on their weak axis when used for story-high walls; a simple requirement deriving from the limiting ratio of L/d of 50 for a solid-sawn column. If the wall is surfaced on both sides, the studs are usually considered to be adequately braced by the surfacing. If the wall is not surfaced, or is surfaced on only one side, blocking between studs must be provided, as shown in Fig. 12-5. The number of rows of

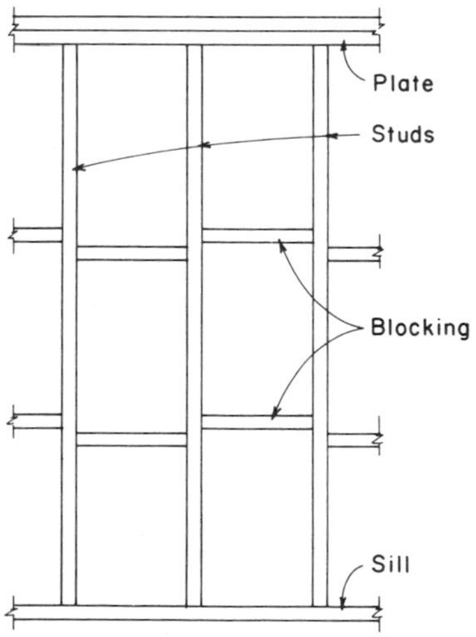

FIGURE 12-5. Stud bracing.

blocking and the actual spacing of the blocking will depend on the wall height and the need for column action by the studs.

Studs may also serve other functions, as in the case of an exterior wall where they must span to resist wind forces on the wall. For this situation the studs must be designed for the combined actions of bending plus compression, as discussed in Sec. 12-10.

In colder climates it is now common to use studs of greater width than the nominal 4 in. in order to create a larger void space within the wall to accommodate insulation. This often results in studs that are quite redundantly strong for ordinary tasks in one- and two-story buildings. Of course, wider studs may also be required for very tall walls; the 2 × 4 is generally limited to a height of 14 ft.

If vertical loads are high or lateral bending is great, it may be necessary to strengthen a stud wall. This can be done in a number of ways such as:

Decreasing stud spacing; from the usual 16 in. to 12 in., for example.

Increasing the stud thickness from 2 in. to 3 in. nominal.

Increasing the stud width (which thickens the wall).

Using doubled or tripled studs (or larger timber sections) as posts at locations of concentrated loads.

It is also sometimes necessary to use thicker studs or to restrict stud spacing for walls that function as shear walls, as discussed in Chapter 19.

In general, studs are columns and must comply to the various requirements for design of solid-sawn sections. Any appropriate grade of wood may be used, although special stud grades are commonly used for ordinary 2 × 4 studs.

12-8 Spaced Columns

A type of structural element sometimes used in wood structures is the *spaced column*. This is an element in which two or more wood members are fastened together to share load as a single compression unit. The design of such elements is quite complex, owing to the numerous code requirements. The following example shows the general procedure for analysis of such an element, but the reader should refer to the applicable code for the various requirements for any design work.

Example. A spaced column of the form shown in Fig. 12-6 consists of three 3 × 10 pieces of Douglas fir–larch, No. 1 Dense

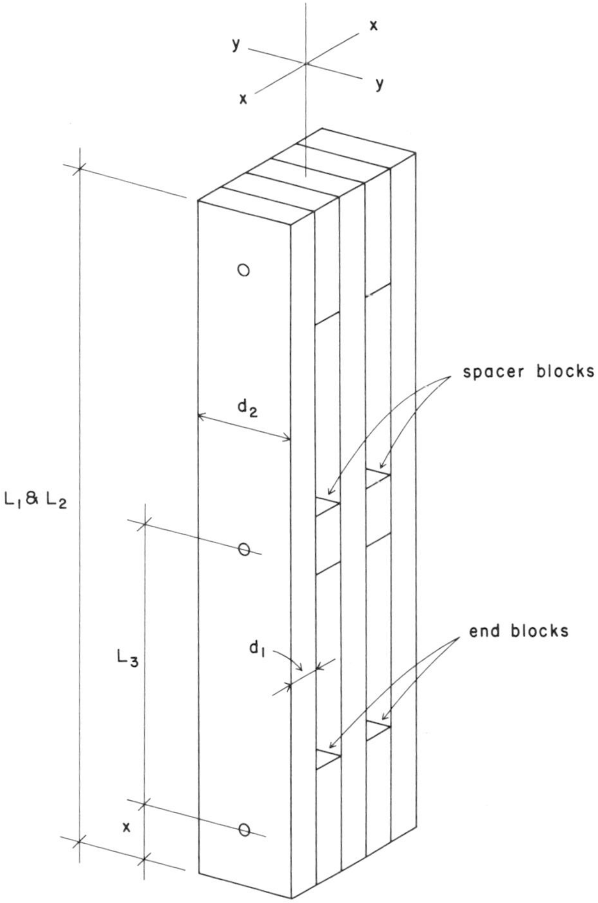

FIGURE 12-6. The spaced column.

grade. Dimension $L_1 = 15$ ft [4.57 m] and $x = 6$ in. [152 mm]. Find the axial compression capacity.

Solution: There are two separate conditions to be investigated for the spaced column. These relate to the effects of relative slenderness in the two directions, as designated by the x- and y-axes in Fig. 12-6. In the y direction the column behaves simply as a set of solid wood columns. Thus the stress permitted is limited by the dimension d_2 and the ratio of L_2 to d_2, and F'_c for this condition is the same as that for a solid wood column. Thus, for the example

$$\frac{L_2}{d_2} = \frac{15 \times 12}{9.25} = 19.46$$

and F'_c is determined as usual for a solid section.

We determine that

$$K = 0.671 \sqrt{\frac{E}{F_c}} = 0.671 \sqrt{\frac{1{,}900{,}000}{1450}} = 24.29$$

using values for E and F_c from Table 13-1.

This establishes the stress condition as zone 2 as shown in Fig. 12-2 (L/d between 11 and K). The allowable stress is thus

$$F'_c = F_c \left\{1 - \frac{1}{3}\left(\frac{L/d}{K}\right)^4\right\}$$

$$= 1450 \left\{1 - \frac{1}{3}\left(\frac{19.46}{24.29}\right)^4\right\} = 1251 \text{ psi } [8.63 \text{ MPa}]$$

For the condition of behavior with regard to the x direction, we first check for conformance with two limitations:

1. Maximum value for $L_3/d_1 = 40$.
2. Maximum value for $L_1/d_1 = 80$.

Thus

$$\frac{84}{2.5} = 33.6 < 40$$

and

$$\frac{180}{2.5} = 72 < 80$$

so the limits are not exceeded.

The stress permitted for the x direction condition depends on the value of L_1/d_1 and is one of three situations, similar to the requirements for solid columns:

1. For values of L_1/d_1 of 11 or less, $F'_c = F_c$.
2. For values of L_1/d_1 between 11 and K,

$$F'_c = F_c \left\{1 - \frac{1}{3}\left(\frac{L_1/d_1}{K}\right)^4\right\}$$

where $K = 0.671 \sqrt{C_x\left(\frac{E}{F_c}\right)}$

3. For values of L_1/d_1 between K and 80,

$$F'_c = \frac{0.30(C_x)(E)}{(L_1/d_1)^2}$$

In the equations for conditions 2 and 3, the value for C_x is based on the conditions of the end blocks in the column. In the illustration in Fig. 12-6, the distance x indicates the distance from the end of the column to the centroid of the connectors that are used to bolt end blocks into the column. Two values are given for C_x, based on the relation of the distance x to the column length—L_1 in the figure:

1. $C_x = 2.5$ when x is equal to or less than $L_1/20$.
2. $C_x = 3.0$ when x is between $L_1/20$ and $L_1/10$.

For our example, with $x = 6$ in., $L_1/20 = 180/20 = 9$. Thus $C_x = 2.5$ and

$$K = 0.671\sqrt{C_x\left(\frac{E}{F_c}\right)} = 0.671\sqrt{(2.5)\left(\frac{1{,}900{,}000}{1450}\right)} = 38.4$$

which indicates that the stress condition is condition 3 and

$$F'_c = \frac{0.30(C_x)(E)}{(L_1/d_1)^2} = \frac{(0.30)(2.5)(1{,}900{,}000)}{(72)^2}$$

$$= 275 \text{ psi } [1.90 \text{ MPa}]$$

We thus establish that the behavior with respect to the x direction is critical for this column, that the stress is limited to 275 psi, and that the load permitted for the column is thus

$$\text{load} = \text{allowable stress} \times \text{gross area of column}$$

$$= 275 \times (3 \times 23.125)$$

$$= 19{,}078 \text{ lb } [84.9 \text{ kN}]$$

Problem 12-8-A. A spaced column of the form shown in Fig. 12-6 consists of two 2 × 8 pieces of Southern pine, No. 1 grade. Dimension $L_1 = 10$ ft [3.05 m] and $x = 5$ in. [127 mm]. Find the axial compression capacity for the column.

12-9 Built-up Columns

In various situations single columns may consist of multiple elements of solid-sawn sections. Although the description includes glued-laminated and spaced columns, the term *built-up* column is generally used for multiple element columns such as those shown in Fig. 12-7. Glued-laminated columns are essentially designed as solid sections. Spaced columns that qualify for the designation are designed as discussed in Sec. 12-8.

Built-up columns generally have the elements attached to each other by mechanical devices such as nails, spikes, lag screws, or machine bolts. Unless the particular assembly has been load tested for code approval, it is usually designed on the basis of the single element capacity. That is, the *least* load capacity of the

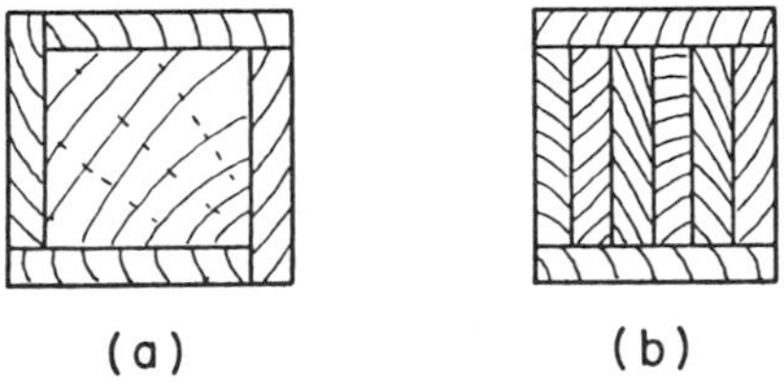

FIGURE 12-7. Built-up columns.

built-up section is the sum of the capacities of the individually considered parts.

The most common built-up columns are the multiple stud assemblies that occur at wall corners, wall intersections, and the edges of door and window openings. When braced by wall surfacing or appropriate blocking, the aggregate capacity of such assemblies is considered to be the sum of the individual stud capacities.

When built-up columns occur as freestanding columns, it may be difficult to make a case for rational determination of their capacities unless single elements have low enough slenderness to qualify for significant capacities. Two types of assemblies that have proven capabilities as built-up columns are those shown in Fig. 12-7. In Fig. 12-7*a* a solid core column is wrapped on all sides by thinner elements. The common assumption for this column is that slenderness is based on the core alone but that axial compression capacity is based on the whole section.

In Fig. 12-7*b* a series of thin elements is held together by two cover plates that tend to restrict the buckling of the core elements on their weak axes. For this column slenderness is considered to be based on the stronger axis of the inner members. Axial compression may be based on the sum of the inner elements for a conservative design, but it is reasonable to include the plates if they are attached by screws or bolts.

12-10 Columns with Bending

There are a number of situations in which structural members are subjected to combined effects of axial compression and bending. Stresses developed by these two actions are both of the direct type (tension and compression) and can be combined for consideration of a net stress condition. However, the basic actions of a column and a bending member are essentially different in character, and it is therefore customary to consider this combined activity by what is called *interaction.*

The classic form of interaction is represented by the graph in Fig. 12-8. Referring to the notation on the graph:

The maximum axial load capacity of the column (without bending) is P_0.

The maximum bending capacity of the member (without compression) is M_0.

At some applied compression load below P_0 the column has some capacity for bending in combination with the load. This combination is indicated as P_n and M_n.

The classic form of the interaction relationship is expressed in the formula

$$\frac{P_n}{P_0} + \frac{M_n}{M_0} = 1$$

The plot of this equation is the straight line connecting P_0 and M_0 as shown on Fig. 12-8.

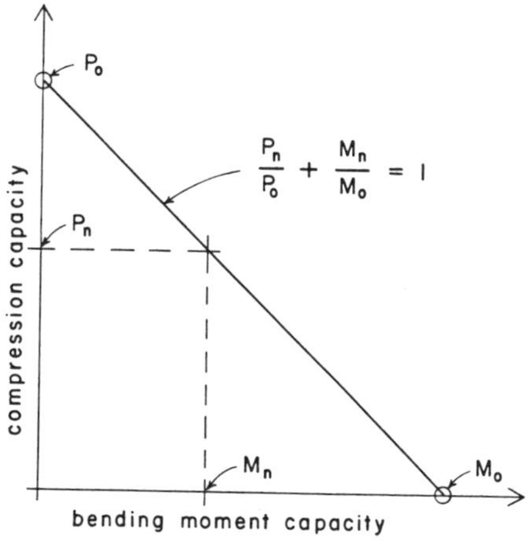

FIGURE 12-8. Column interaction: compression plus bending.

A graph similar to that in Fig. 12-8 can be produced using stresses rather than loads and moments. This is the procedure generally used in wood and steel design, with the graph taking a form expressed as

$$\frac{f_a}{F_a} + \frac{f_b}{F_b} \leq 1$$

in which f_a = computed stress due to the actual load,
F_a = allowable column action stress,
f_b = computed stress due to bending,
F_b = allowable bending stress.

Various effects cause deviations from the pure, straight line form of interaction, including inelastic behavior, effects of lateral stability or torsion, and general effects of member cross section form. A major effect is the so-called P-delta effect. Figure 12-9a

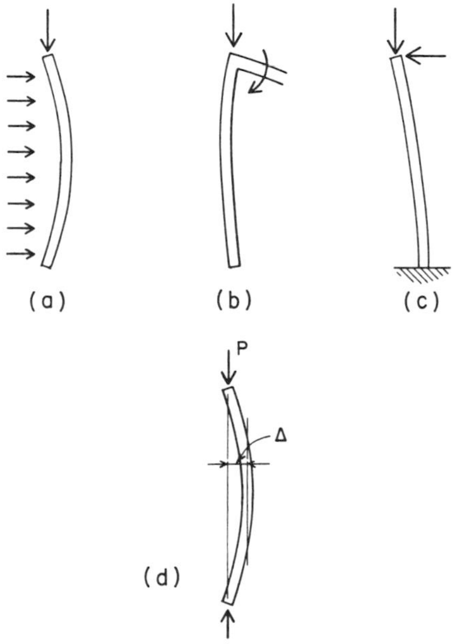

FIGURE 12-9. The P-delta effect.

shows a common situation that occurs in buildings when an exterior wall functions as a bearing wall or contains a column. The combination of gravity and lateral load due to wind or seismic action can produce the loading condition shown. If the member is quite flexible and the deflection is significant in magnitude, an additional moment is developed as the axis of the member deviates from the action line of the vertical compression load. The resulting additional moment is the product of the load (P) and the deflection (Δ), yielding the term used to describe the phenomenon. (See Fig. 12-9*d*.)

Various other situations can result in the P-delta effect. Figure 12-9*b* shows an end column in a rigid frame structure, where moment is induced at the top of the column by the moment-resistive connection to the beam. Although the form of deflection is different in this case, the P-delta effect is similar. The vertically cantilevered column in Fig. 12-9*c* presents potentially an extreme case of this effect.

The P-delta effect may or may not be a critical concern; a major factor is the relative slenderness and flexibility of the column. Relatively stiff columns will both tolerate the eccentric load effect and sustain little deflection due to bending, making the P-delta effect quite insignificant. In a worst-case scenario, however, the P-delta effect can be an accelerating one in which the added moment due to the P-delta effect causes additional deflection, which in turn results in additional P-delta effect, which then causes more deflection, and so on. Potentially critical situations are those involving very slender compression members for which the phenomenon should be carefully considered.

In wood structures, columns with bending occur most frequently as shown in Fig. 12-10. Studs in exterior walls represent the situation shown in Fig. 12-10*a*. In various situations, due to framing details, a column carrying only vertical loads may sustain bending if the load is not axial to the column, as shown in Fig. 12-10*b*. Current criteria for wood column design use the simple straight line interaction relationship as a fundamental reference with various adjustments for special cases. One adjustment is for the P-delta effect just described. Another adjustment relates to the potential for buckling due to bending, which is treated by

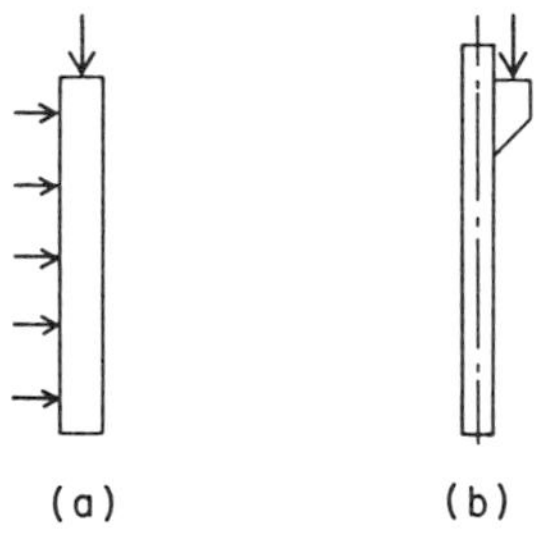

FIGURE 12-10. Development of combined compression and bending in columns—common cases.

modifications of the allowable bending stress. The general form of the interaction relationship is

$$\frac{f_c}{F'_c} + \frac{f_b}{F'_b - Jf_c} \leq 1$$

In this formula f_c and f_b are the computed axial compression stress and bending stress, respectively. F'_c is the usual allowable column compressive stress, as discussed in Sec. 12-3. F'_b is the allowable bending stress that is adjusted if necessary for any stability considerations. J is a factor determined from the equation

$$J = \frac{(L_e/d) - 11}{K - 11}$$

in which L_e = the unbraced height in the plane of bending,
d = the column width in the plane of bending,
K = the limiting slenderness ratio as used in the usual determination of F'_c.

For the three zones of relative column stiffness, as shown in Fig. 12-2, the use of J is as follows:

Zone 1. $L_e/d \leq 11$, $J = 0$.

Zone 2. $11 \leq L_e/d \leq K$, J = the computed value from the preceding formula.

Zone 3. $L_e \geq K$, $J = 1$.

For slender, unbraced columns, it may be necessary to use an adjusted value for F_b. However, based on code qualifications, the following exceptions are noted:

1. For square sections used as beams, no lateral support is required and no stress adjustment is made.
2. When the compression edge is continuously supported (as for a typical wall stud), the unsupported length may be considered zero (i.e., the laterally unsupported length) and no adjustment of bending stress is required.

For these two cases, $F'_b = F_b$ (from Table 3-1), although the adjustment for J must still be considered.

The following examples illustrate the use of these relationships. Some additional examples are given in the computations in Chapter 20.

Example 1. An exterior wall stud is loaded as shown in Fig. 12-11. Studs are 12 ft high and are of Douglas fir–larch, Stud grade. Investigate the stud for combined bending and column action.

Solution: From Table 3-1 we note that F_b = 850 psi (repetitive member use), F_c = 600 psi, E = 1,500,000 psi. With the inclusion of the wind loading, these may be increased by one-third (see Sec. 3-3).

We assume that the wall surfacing materials provide adequate bracing for the studs on their weak axis (d = 1.5 in.), so that the dimension to be used for both the allowable column stress and bending stress reduction is the other dimension, that is, 5.5 in.

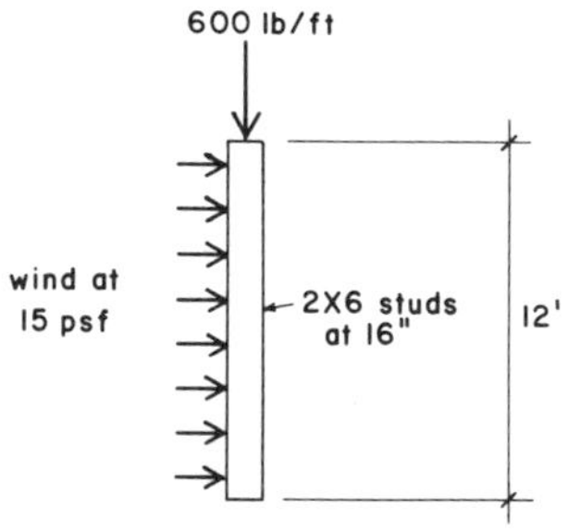

FIGURE 12-11.

Thus

$$\frac{L}{d} = \frac{(12 \times 12)}{5.5} = 26.18$$

$$K = 0.671\sqrt{\frac{E}{F_c}} = 0.671\sqrt{\frac{1{,}500{,}000}{600}} = 33.55$$

$$F'_c = F_c\left[1 - \frac{1}{3}\left(\frac{L/d}{K}\right)^4\right] = 600\left[1 - \frac{1}{3}\left(\frac{26.18}{33.55}\right)^4\right] = 526 \text{ psi}$$

$$\frac{L_e}{d} = \frac{(12 \times 12)}{5.5} = 26.18$$

$$J = \frac{(L_e/d) - 11}{K - 11} = \frac{26.18 - 11}{33.55 - 11} = 0.673$$

For the computed stresses:

$$f_c = \frac{P}{A} = \frac{(16/12)(600)}{8.25} = 97 \text{ psi}$$

$$M = \frac{16}{12}\,\frac{(15)(12)^2}{8} = 360 \text{ ft-lb}$$

$$f_b = \frac{M}{S} = \frac{(360 \times 12)}{7.563} = 571 \text{ psi}$$

With the stud fully braced on its weak axis, the value for F'_b is taken as the unreduced value from Table 3-1. Thus the value used in the interaction formula is

$$F'_b - Jf_c = 850 - (0.673 \times 97) = 785 \text{ psi}$$

The interaction formula is thus

$$\frac{f_c}{F'_c} + \frac{f_b}{F'_b - Jf_c} = \frac{97}{1.33(526)} + \frac{571}{1.33(785)} = 0.139 + 0.547$$

$$= 0.686$$

As this is less than one, the stud is adequate.

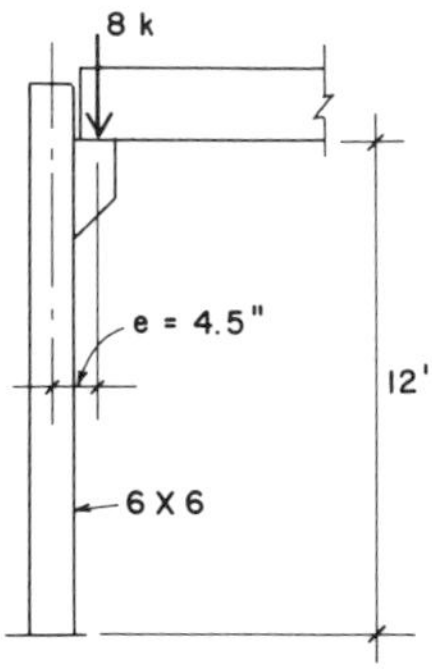

FIGURE 12-12.

Example 2. The column shown in Fig. 12-12 is of Douglas fir–larch, Dense No. 1 grade. Investigate the column for combined column and bending actions.

Solution: From Table 3-1, $F_b = 1400$ psi, $F_c = 1200$ psi, and $E = 1{,}700{,}000$ psi. From Table 4-1, $A = 30.25$ in.^{2}, $S = 27.7$ in.^{3}. Then

$$\frac{L}{d} = \frac{(12 \times 12)}{5.5} = 26.18$$

$$K = 0.671\sqrt{\frac{1{,}700{,}000}{1200}} = 25.25$$

$$F_c' = \frac{0.3E}{(L/d)^2} = \frac{0.3(1{,}700{,}000)}{(26.18)^2} = 744 \text{ psi}$$

$$F_b' = F_b = 1400 \text{ psi}$$

$$f_c = \frac{8000}{30.25} = 264 \text{ psi}$$

$$f_b = \frac{(8000 \times 4.5)}{27.7} = 1300 \text{ psi}$$

$$J = 1$$

Thus

$$\frac{f_c}{F'_c} + \frac{f_b}{F'_b - Jf_c} = \frac{264}{744} + \frac{1300}{1400 - 264} = 0.355 + 1.14 = 1.495$$

As this exceeds one, the column is not adequate.

Problem 12-10-A. Nine feet high 2 × 4 studs of Douglas fir–larch, No. 1 grade, are used in an exterior wall. Wind load is 17 psf on the wall surface; studs are 24 in. on center; the gravity load on the wall is 400 lb/ft of wall length. Investigate the studs for combined action of compression and bending.

Problem 12-10-B. A 10 × 10 column of Douglas fir–larch, No. 1 grade, is 9 ft high and carries a compression load that is 7.5 in. eccentric from the column axis. Investigate the column for combined compression and bending.

13

Wood Fastenings

Assembling wood structures involves a great amount of joining and connecting of individual pieces. Various types of fastening devices are used in this assemblage. Some of the common means for attachment are considered in this chapter. Some special connection devices and problems are considered in Chapter 14.

13-1 Bolted Joints in Wood Structures

When steel bolts are used to connect wood members, there are several design considerations. The principal concerns are the following:

1. *Net Stress in Member.* Holes drilled for the placing of bolts reduce the member cross section. For this analysis the hole is assumed to have a diameter $\frac{1}{16}$ in. larger than that of the bolt. The most common situations are those shown in Fig. 13-1. When bolts are staggered, it may be necessary to make two investigations, as shown in the illustration.
2. *Bearing of the Bolt on the Wood and Bending in the Bolt.* When the members are thick and the bolt thin and long, the

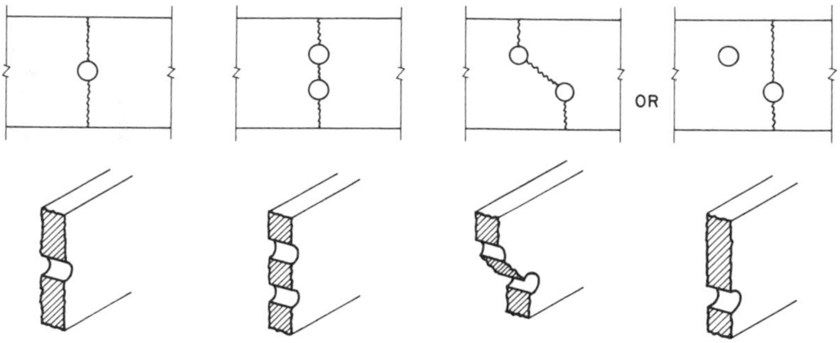

FIGURE 13-1. Effect of bolt holes on reduction of cross section for tension members.

bending of the bolt will cause a concentration of stress at the edges of the members. The bearing on the wood is further limited by the angle of the load to the grain because wood is much stronger in compression in the grain direction.

3. *Number of Members Bolted.* The worst case, as shown in Fig. 13-2, is that of the two-member joint. In this case the lack of symmetry in the joint produces considerable twisting. This situation is referred to as single shear since the bolt is subjected to shear on a single plane. When more members are joined, this twisting effect is reduced.
4. *Ripping Out the Bolt When Too Close to an Edge.* This problem, together with that of the minimum spacing of the bolts in multiple-bolt joints, is dealt with by using the criteria given in Fig. 13-3. Note that the limiting dimensions involve the consideration of the bolt diameter *D;* the bolt length *L;* the type of force—tension or compression; and the angle of load to the grain of the wood.

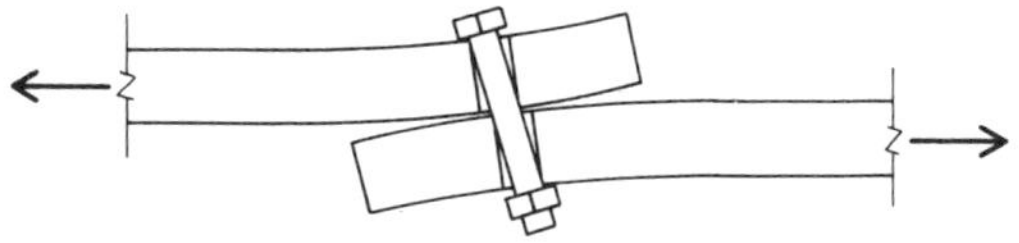

FIGURE 13-2. Behavior of the single lapped joint.

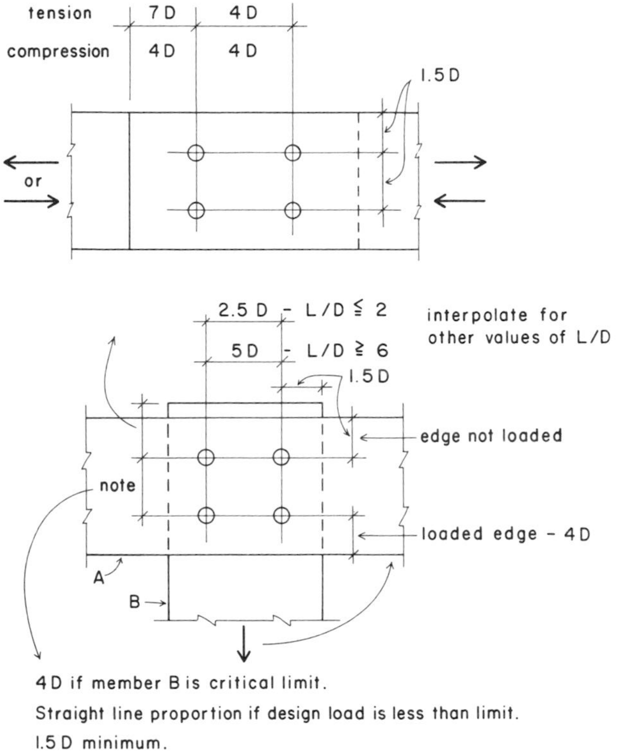

FIGURE 13-3. Edge, end, and spacing distances for bolts in wood structures.

The bolt design length is established on the basis of the number of members in the joint and the thickness of the wood pieces. There are many possible cases, but the most common are those shown in Fig. 13-4. The critical lengths for these cases are given in Table 13-1. Also given in the table is the factor for determining the allowable load on the bolt. Allowable loads ordinarily are tabulated for the three-member joint (Case 1 in Fig. 13-4), and the factors represent adjustments for other conditions.

Table 13-2 gives allowable loads for bolts with wood members of dense grades of Douglas fir–larch and Southern pine. The two

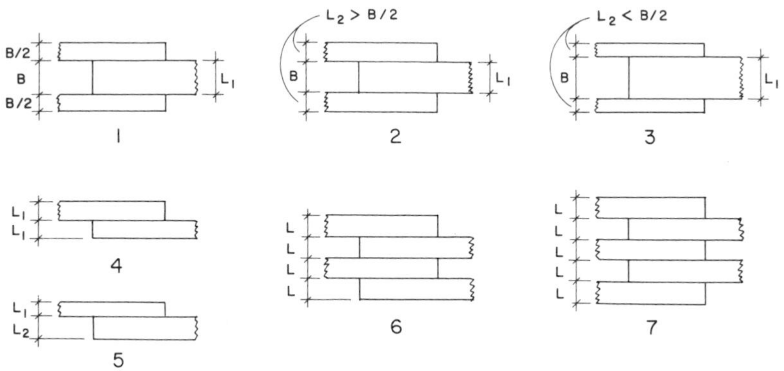

FIGURE 13-4. Various cases of lapped joints with relation to the determination of the critical bolt length.

loads given are that for a load parallel to the grain (*P* load) and that for a load perpendicular to the grain (*Q* load). Figure 13-5 illustrates these two loading conditions, together with the case of a load at some other angle (Θ). For such cases it is necessary to find the allowable load for the specific angle. Figure 13-6 is an adaptation of the Hankinson graph, which may be used to find values for loads at some angle to the grain.

The following examples illustrate the use of the data presented for the design of bolted joints.

Example 1. A three-member bolted joint is made with members of Douglas fir–larch Dense No. 1 grade. The members are loaded

TABLE 13-1. Design Length for Bolts

Case[a]	Critical Length	Modification Factor
1	L_1	1.0
2	L_1	1.0
3	$2L_2$	1.0
4	L_1	0.5
5	Lesser of L_2 or $2L_1$	0.5
6	L	1.5
7	L	2.0

[a] See Figure 13-4.

TABLE 13-2. Bolt Design Values for Wood Joints for Douglas Fir–Larch and Southern Pine[a]

Design Length of Bolt (in.)	Diameter of Bolt (in.)	Design Values for One Bolt in Double Shear[a] (lb)			
		Parallel to Grain Load (P)		Perpendicular to Grain Load (Q)	
		Dense Grades	Ordinary Grades	Dense Grades	Ordinary Grades
1.5	1/2	1100	940	500	430
	5/8	1380	1180	570	490
	3/4	1660	1420	630	540
	7/8	1940	1660	700	600
	1	2220	1890	760	650
2.5	1/2	1480	1260	840	720
	5/8	2140	1820	950	810
	3/4	2710	2310	1060	900
	7/8	3210	2740	1160	990
	1	3680	3150	1270	1080
3.0	1/2	1490	1270	1010	860
	5/8	2290	1960	1140	970
	3/4	3080	2630	1270	1080
	7/8	3770	3220	1390	1190
	1	4390	3750	1520	1300
3.5	1/2	1490	1270	1140	980
	5/8	2320	1980	1330	1130
	3/4	3280	2800	1480	1260
	7/8	4190	3580	1630	1390
	1	5000	4270	1770	1520
5.5	5/8	2330	1990	1650	1410
	3/4	3350	2860	2200	1880
	7/8	4570	3900	2550	2180
	1	5930	5070	2790	2380
	1 1/4	8940	7640	3260	2790
7.5	5/8	2330	1990	1480	1260
	3/4	3350	2860	2130	1820
	7/8	4560	3890	2840	2430
	1	5950	5080	3550	3030
	1 1/4	9310	7950	4450	3800

[a] See Table 13-1 for modification factors for other conditions.

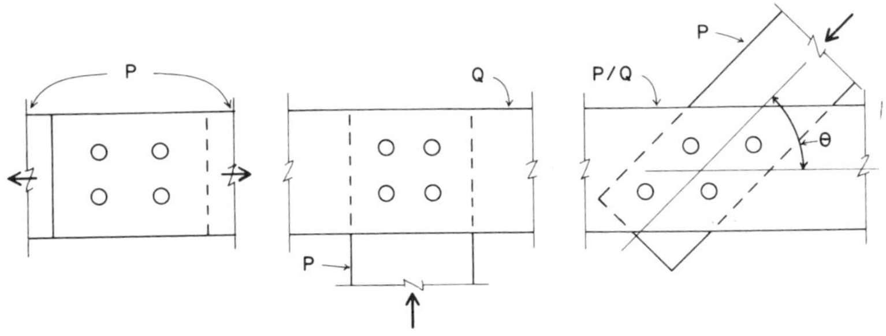

(a) (b) (c)

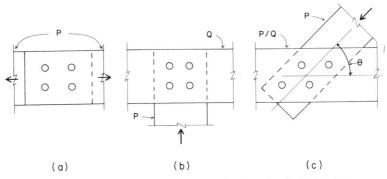

FIGURE 13-5. Relation of load to grain direction in bolted joints.

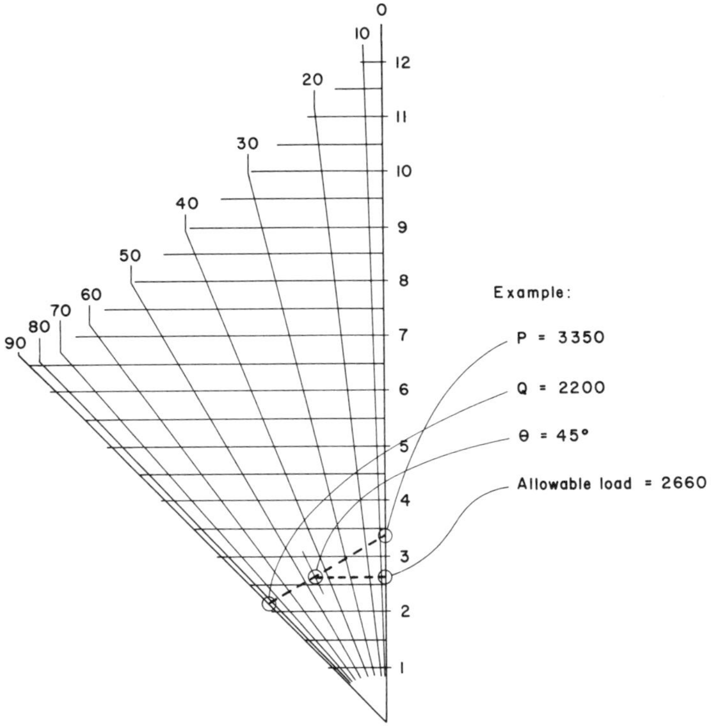

FIGURE 13-6. Hankinson graph for determination of load values with the loading at an angle to the wood grain.

in a direction parallel to the grain (Fig. 13-5*a*) with a tension force of 10 kips. The middle member is a 3 × 12 and the outer members are each 2 × 12. If four $\frac{3}{4}$ in. bolts are used for the joint, is the joint capable of carrying the tension load?

Solution: The first step is to identify the critical bolt length (Fig. 13-4) and the load factor (Table 13-1). Since the outer members are greater than one-half the thickness of the middle member, the condition is that of Case 2 in Fig. 13-4, and the effective length is the thickness of the middle member, 2.5 in. The load factor from Table 13-1 is 1.0, indicating that the tabulated load may be used with no adjustment.

From Table 13-2 the allowable load per bolt is 2710 lb. (Bolt design length of 2.5 in.; bolt diameter of $\frac{3}{4}$ in.; *P* load; dense grade.) With the four bolts, the total capacity of the bolts is thus 4 × 2710 = 10,840 lb.

For tension stress in the wood, the critical condition is for the middle member, with the net section through the bolts being that shown in Fig. 13-1*b*. With the holes being considered as $\frac{1}{16}$ in. larger than the bolts, the net area for tension stress is thus

$$A = 2.5 \times \left[11.25 - \left(2 \times \frac{13}{16}\right)\right] = 24.1 \text{ in.}^2$$

From Table 3-1 the allowable tension stress is 1200 psi. The maximum tension capacity of the member at the net section is thus

$$T = (\text{allowable stress}) \times (\text{net area}) = 1200 \times 24.1 = 28{,}900 \text{ lb}$$

and the joint is adequate for the required load.

Example 2. A bolted two-member joint consists of two 2 × 10 members of Douglas fir–larch, Dense No. 2 grade, attached at right angles to each other, as shown in Fig. 13–5*b*. If the joint is made with two $\frac{7}{8}$ in. bolts, what is the maximum compression capacity of the joint?

Solution: This is Case 4 in Fig. 13-4, and the effective length is the member thickness of 1.5 in. The modification factor from

Table 13-1 is 0.5, and the bolt capacity from Table 13-2 is 700 lb per bolt. (Bolt design length of 1.5 in.; bolt diameter of $\frac{7}{8}$ in.; Q load; dense grade.) The total capacity of the bolts is thus

$$C = 2 \times 700 \times 0.5 = 700 \text{ lb}$$

The net section is not a concern for the compression force, as the capacity of the members would be based on an analysis for the slenderness condition based on the L/d of the members.

Example 3. A three-member bolted joint consists of two outer members, each 2 × 10, and the middle member that is a 4 × 12. The outer members are arranged at an angle to the middle member, as shown in Fig. 13-5*c*, such that $\Theta = 60°$. Find the maximum compression force that can be transmitted through the joint by the outer members. Wood is Southern pine, No. 1 grade. The joint is made with two $\frac{3}{4}$-in. bolts.

Solution: In this case we must investigate both the outer and middle members. For the outer members the effective length is 2 × 1.5 = 3.0 in., and the modification factor is 1.0 (Case 3, Fig. 13-4 and Table 13-1). From Table 13-2 the bolt capacity based on the outer members is 2630 lb per bolt. (Bolt design length of 3.0 in.; bolt diameter of $\frac{3}{4}$ in.; P load; ordinary grade.)

For the middle member the effective bolt length is the member thickness of 3.5 in., and the unadjusted load per bolt from Table 13-2 is 2800 lb for the P condition and 1260 lb for the Q condition. If these values are used on the Hankinson graph in Fig. 13-6, the load per bolt for the 60° angle is found to be approximately 1700 lb. Since this value is lower than that found for the outer members, it represents the limit for the joint. The joint capacity based on the bolts is thus 2 × 1700 = 3400 lb.

(*Note:* For all of the following problems, use Douglas fir–larch, Dense No. 1 grade.)

Problem 13-1-A*. A three-member tension joint has 2 × 12 outer members and a 4 × 12 middle member (Fig. 13-5*a*). The joint is made with six $\frac{3}{4}$ in. bolts in two rows. Find the capacity of the joint as limited by the bolts and the tension stresses in the members.

Problem 13-1-B. A two-member tension joint consists of 2 × 6 members bolted with two $\frac{7}{8}$ in. bolts (Fig. 13-5*a*). What is the limit for the tension force?

Problem 13-1-C. Two outer members, each 2 × 8, are bolted to a middle member consisting of a 3 × 12 with two 1 in. bolts. The outer members form an angle of 45° with respect to the middle member (Fig. 13-5*c*; $\Theta = 45°$). What is the maximum compression force that the outer members can transmit to the joint?

13-2 Nailed Joints

Nails are made in a wide range of sizes and forms for many purposes. They range in size from tiny tacks to huge spikes. Most nails are driven by someone pounding a hammer—more or less as they have been for thousands of years. For situations where many nails must be driven, however, we now have a variety of mechanical driving devices. Use of manual and powered mechanical driving equipment has resulted in other types of fasteners, such as staples and screws, thus replacing nails in some types of connections. Mechanically installed devices are discussed in general in Sec. 13-4.

For structural fastening in light wood framing, the nail most commonly used is called—appropriately—the *common wire nail,* or simply the common nail. Basic concerns for the use of the common nail, as shown in Fig. 13-7, are the following:

1. *Nail Size.* Critical dimensions are the diameter and length. Sizes are specified in pennyweight units, designated as 4d, 6d, and so on, and referred to as four penny, six penny, and so on.

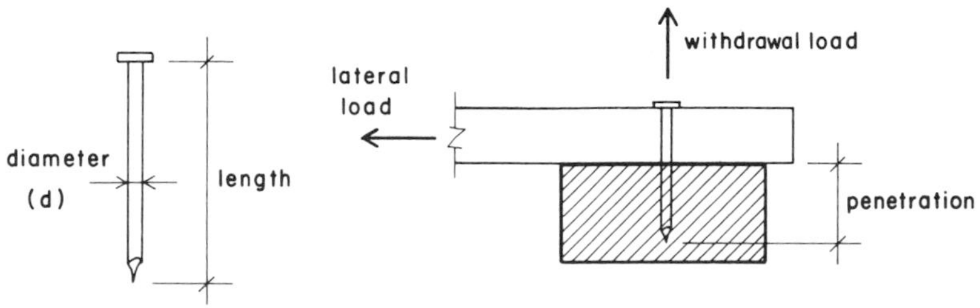

FIGURE 13-7. Typical common wire nail and loading conditions.

2. *Load Direction.* Pull-out loading in the direction of the nail shaft is called *withdrawal;* shear loading perpendicular to the nail shaft is called *lateral load.*
3. *Penetration.* Nailing is typically done through one element and into another, and the load capacity is limited by the amount of length of embedment of the nail in the second member. The length of embedment is called the *penetration.*
4. *Species and Grade of Wood.* The harder, tougher, and heavier the wood, the more the load resistance capability.

Design of good nail joints requires a little engineering and a lot of good carpentry. Some obvious situations to avoid are those shown in Fig. 13-8. A little actual carpentry experience is highly desirable for anyone who designs nailed joints.

Withdrawal load capacities of common wire nails are given in Table 13-3. The capacities are given for both Douglas fir–larch and Southern pine. Note that the table values are given in units of capacity per inch of penetration and must be multiplied by the actual penetration length to obtain the load capacity in pounds. In general, it is best not to use structural joints that rely on withdrawal loading resistance.

Lateral load capacities for common wire nails are given in Table 13-4. These values also apply to Douglas fir–larch and Southern pine. Note that a penetration of at least 11 times the nail diameter is required for the development of the full capacity of the nail. A value of one-third of that in the table is permitted with a penetration of one-third of this length, which is the minimum

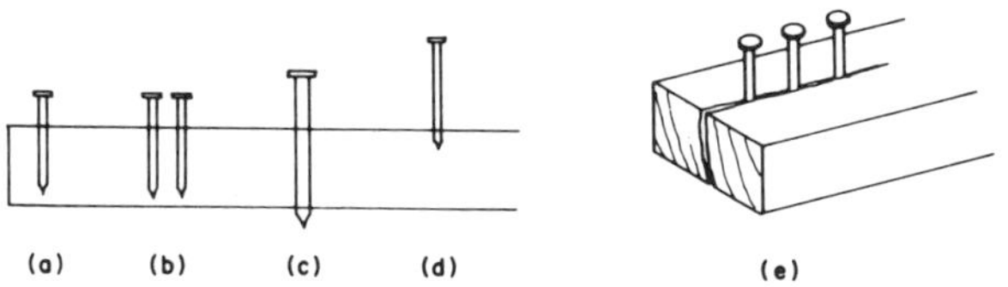

FIGURE 13-8. Poor nailing practices: (*a*) too close to edge; (*b*) nails too close; (*c*) nail too large for wood piece; (*d*) too little penetration in holding piece; (*e*) too many nails in a single row parallel to wood grain.

TABLE 13-3. Withdrawal Load Capacity of Common Wire Nails (lb/in.)

	Size of Nail				
Pennyweight	6	8	10	12	16
Diameter (in.)	0.113	0.131	0.148	0.148	0.162
Douglas fir–larch	29	34	38	38	42
Southern pine	35	41	46	46	50

penetration permitted. For actual penetration lengths between these limits, the load capacity may be determined by direct proportion. Orientation of the load to the direction of grain in the wood is not a concern when considering nails in terms of lateral loading.

The following example illustrates the design of a typical nailed joint for a wood truss.

Example. The truss heel joint shown in Fig. 13-9 is made with 2 in. nominal wood elements of Douglas fir–larch, Dense No. 1 grade, and gusset plates of $\frac{1}{2}$ in. plywood. Nails are 6d common, with the nail layout shown occurring on both sides of the joint. Find the tension force limit for the bottom chord (load 3 in the illustration).

Solution: The two primary concerns are for the lateral capacity of the nails and the tension, tearing stress in the gussets. For the nails, we observe from Table 13-4 that

TABLE 13-4. Lateral Load Capacity of Common Wire Nails (lb/nail)

	Size of Nail				
Pennyweight	6	8	10	12	16
Diameter (in.)	0.113	0.131	0.148	0.148	0.162
Length (in.)	2.0	2.5	3.0	3.25	3.5
Douglas fir–larch and Southern pine	63	78	94	94	108
Penetration required for 100% of table value[a] (in.)	1.24	1.44	1.63	1.63	1.78
Minimum penetration[b] (in.)	0.42	0.48	0.54	0.54	0.59

[a] Eleven diameters; reduce by straight-line proportion for less penetration.
[b] One-third of that for full value; 11/3 diameters.

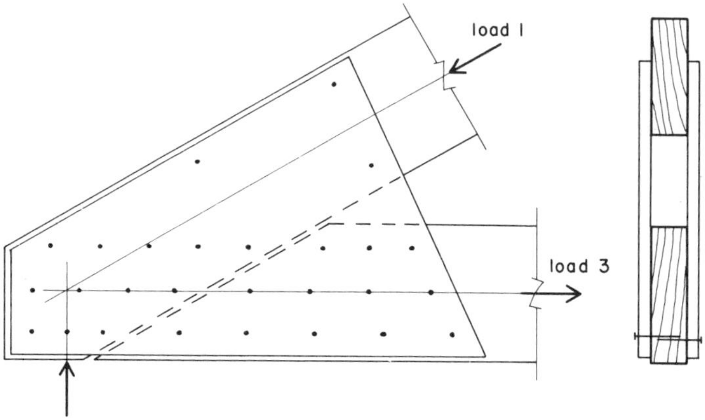

FIGURE 13-9. Truss joint with nails and plywood gusset plates.

Nail length is 2 in. [51 mm].
Minimum penetration for full capacity is 1.24 in. [31.5 mm].
Maximum capacity is 63 lb/nail [0.28 kN].

From inspection of the joint layout, we see that

$$\begin{aligned}\text{actual penetration} &= \text{nail length} - \text{plywood thickness}\\ &= 2.0 - 0.5 = 1.5 \text{ in. [38 mm]}\end{aligned}$$

Therefore we may use the full table value for the nails. With 12 nails on each side of the member, the total capacity is thus

$$F_3 = (24)(63) = 1512 \text{ lb [6.73 kN]}$$

If we consider the cross section of the plywood gussets only in the zone of the bottom chord member, the tension stress in the plywood will be approximately

$$f_t = \frac{1512}{(0.5)(2)(5 \text{ in. of width})} = 302 \text{ psi [2.08 MPa]}$$

which is probably not a critical magnitude for the plywood.

A problem that must be considered in this type of joint is that of the pattern of placement of the nails (commonly called the layout of the nails). To accommodate the large number of nails required, they must be quite closely spaced, and since they are close to the ends of the wood pieces, the possibility of splitting the wood is a critical concern. The factors that determine this possibility include the size of the nail (essentially its diameter), the spacing of the nails, the distance of the nails from the end of the piece, and the tendency of the particular wood species to be split. There are no formal guidelines for this problem; it is largely a matter of good carpentry or some experimentation to establish the feasibility of a given layout.

One technique that can be used to reduce the possibility of splitting is to stagger the nails rather than to arrange them in single rows. Another technique is to use a single set of nails for both gusset plates rather than to nail the plates independently, as shown in Fig. 13-10. The latter procedure consists simply of driving a nail of sufficient length so that its end protrudes from the gusset on the opposite side and then bending the end over—called *clinching*—so that the nail is anchored on both ends. A single nail may thus be utilized for twice its rated capacity for lateral load. This is similar to the development of a single bolt in double shear in a three-member joint.

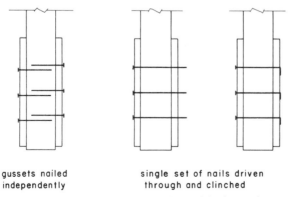

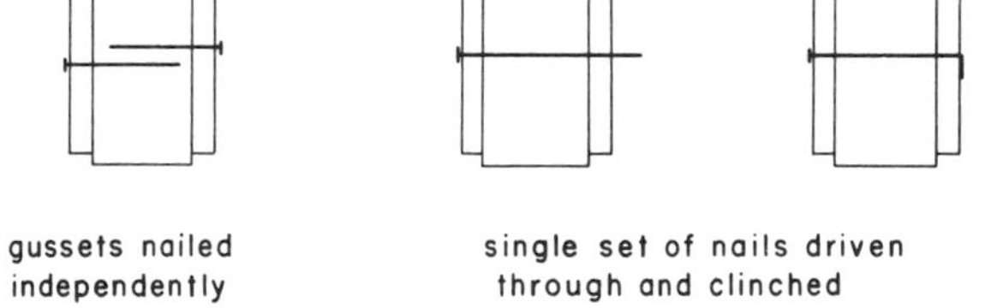

FIGURE 13-10. Nailing techniques for joints with plywood gusset plates.

It is also possible to glue the gusset plates to the wood pieces and to use the nails essentially to hold the plates in place only until the glue has set and hardened. The adequacy of such joints should be verified by load testing, and the nails should be capable of developing some significant percentage of the design load as a safety backup for the glue.

Problem 13-2-A. Two wood members of 3 in. nominal thickness are attached at right angles to each other by plywood gussets on both sides to develop a tension force through the joint. The wood is Douglas fir–larch, No. 1 grade; the plywood gussets are $\frac{3}{4}$ in. thick; nails are common wire. Select the size of nail and find the number required if the tension force is 2000 lb [8.90 kN].

13-3 Screws and Lag Bolts

When loosening or popping of nails is a problem, a more positive grabbing of the nail by the surrounding wood can be achieved by having some nonsmooth surface on the nail shaft. Many special nails with formed surfaces are used, but another means for achieving this effect is to use threaded screws in place of nails. Screws can be tightened down to squeeze connected members together in a manner that is not usually possible with nails. For dynamic loading, such as shaking by an earthquake, the tight, positively anchored connection is a distinct advantage.

Screws are produced in great variety, although three of the most common types used for wood structures are those shown in Fig. 13-11. The flat head screw is designed to be driven so that the head sinks entirely into the wood, resulting in a surface with no protrusions. The round head screw is usually used with a washer, or may be used to attach metal objects to wood surfaces. The hex head screw is called a *lag screw* or *lag bolt* and is designed to be tightened by a wrench rather than by a screwdriver. Lag bolts are made in a considerable size range and can be used for some major structural connections.

Screws function essentially the same as nails, being used to resist either withdrawal or lateral shear-type loading. Screws must usually be installed by first drilling a guide hole, called a *pilot hole,* with a diameter slightly smaller than that of the screw shaft. Specifications for structural connections with screws give

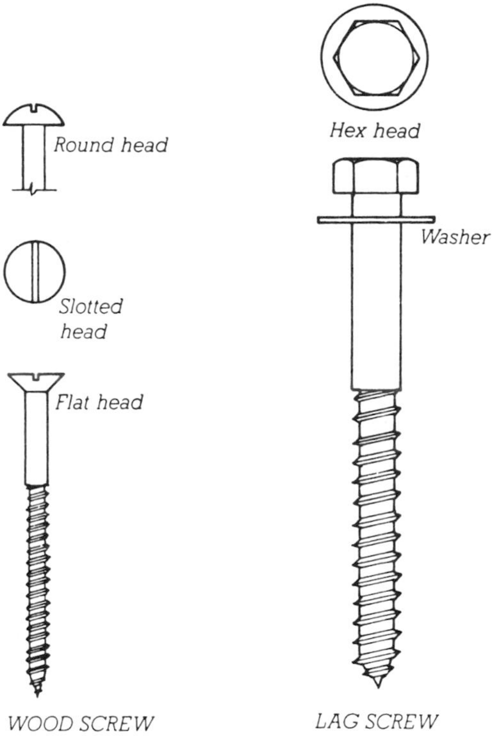

FIGURE 13-11. Types of wood screws. (Reproduced from *The Professional Handbook of Building Construction* by Edward Allen, with permission of the publishers, John Wiley.)

requirements for the limit of pilot holes and for various other details for proper installation. Capacities for withdrawal and lateral loading are given as a function of the screw size and the type of wood holding the screw.

As with nailed joints, the use of screws involves much judgment that is more craft than science. Choice of the screw type, size, spacing, length, and other details of a good joint may be controlled by some specifications, but it is also a matter of experience.

Although it is generally not recommended that nails be relied on for computed loading in withdrawal, screws are not so limited and are often chosen where the details of the connection require such a loading.

13-4 Mechanically Driven Fasteners

Although the hammer, screwdriver, and hand wrench are still in every carpenter's tool kit, many structural fastenings are now routinely achieved with powered devices. In some cases this has produced fasteners that do not fit the old classifications, and codes will eventually probably be more complex to cover the range of fasteners in use. The simple hand staple gun has been extrapolated into a range of devices, including some that can install some significantly strong structural fasteners. Field attachment of plywood roof and floor deck and various wall covering materials is now often accomplished with mechanical driving equipment.

14

Framing Devices

The basic task of attachment is generally achieved by one of the means described in Chapter 13. For various purposes, however, structural connections often utilize some type of intermediate device between wood members and the objects to which they are attached. This chapter presents material relating to some common devices and methods used in producing structural frameworks of wood.

14-1 Shear Developers

When wood members are lapped at a joint and bolted, it is often difficult to prevent some joint movement in the form of slipping between the lapped members. If force reversals, such as those that occur with wind or seismic effects, cause back–and–forth stress on the joint, this lack of tightness in the connection may be objectionable. Various types of devices are sometimes inserted between the lapped members so that when the bolts are tightened, there is some form of resistance to slipping besides the simple friction between the lapped members.

Toothed or ridged devices of metal are sometimes used for this development of enhanced shear resistance in the lapped joint.

They are simply placed between the lapped wood pieces and the tightening of the bolts causes them to bite into both members. Hardware products of various forms and sizes are available and are commonly used, most notably used for heavy, rough timber construction.

A slightly more sophisticated shear developer for the lapped joint is the split-ring connector consisting of a steel ring that is installed by cutting matching circular grooves in the faces of the lapped pieces of wood. When the ring is inserted in the grooves and the bolt is tightened, the ring is squeezed tightly into the grooves and the resulting connection has a shear resistance considerably greater than that with the bolt alone. Design of split-ring connectors is discussed in the next section.

14-2 Split-Ring Connectors

The ordinary form of the split-ring connector and the method of installation are shown in Fig. 14-1. Design considerations for this device include the following:

1. *Size of the Ring*. Rings are available in the two sizes shown in the figure with nominal diameters of 2.5 and 4 in.

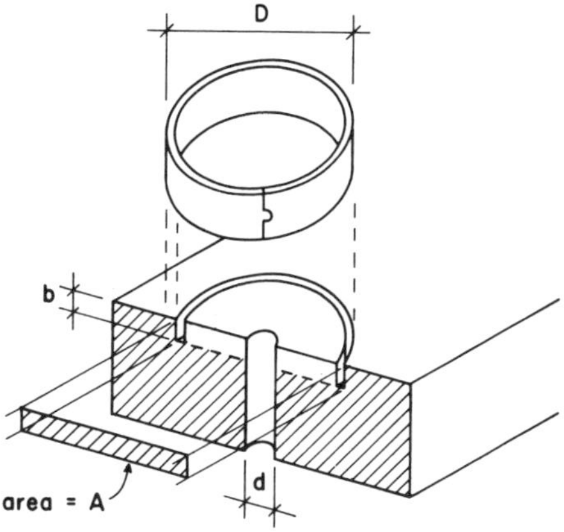

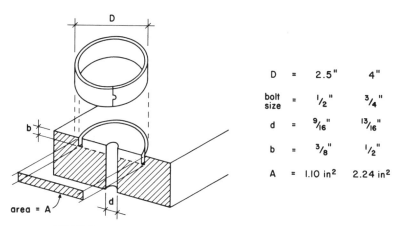

D	=	2.5"	4"
bolt size	=	1/2"	3/4"
d	=	9/16"	13/16"
b	=	3/8"	1/2"
A	=	1.10 in^2	2.24 in^2

FIGURE 14-1. Split-ring connectors for bolted wood joints.

2. *Stress on the Net Section of the Wood Member.* As shown in Fig. 14-1, the cross section of the wood piece is reduced by the ring profile (*A* in the figure) and the bolt hole. If rings are placed on both sides of a wood piece, there will be two reductions for the ring profile.
3. *Thickness of the Wood Piece.* If the wood piece is too thin, the cut for the ring will bite excessively into the cross section. Rated load values reflect concern for this.
4. *Number of Faces of the Wood Piece Having Rings.* As shown in Fig. 14-2, the outside members in a joint will have rings on only one face, whereas the inside members will have rings on both faces. Thickness considerations therefore are more critical for the inside members.
5. *Edge and End Distances.* These must be sufficient to permit the placing of the rings and to prevent splitting out from the side of the wood piece when the joint is loaded. Concern is greatest for the edge in the direction of loading—called the loaded edge. (See Fig. 14-3.)
6. *Spacing of Rings.* Spacing must be sufficient to permit the placing of the rings and the full development of the ring capacity of the wood piece.

Figure 14-4 shows the four placement dimensions that must be considered. The limits for these dimensions are given in Table 14-1. In some cases, two limits are given. One limit is that re-

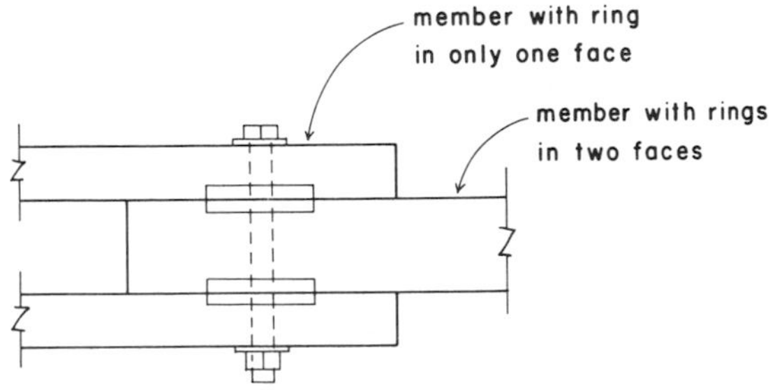

FIGURE 14-2. Determination of the number of faces of a member with split-ring connectors.

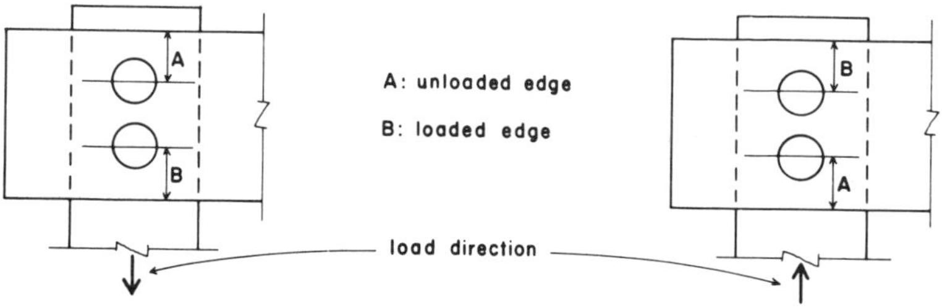

FIGURE 14-3. Determination of the loaded edge condition.

quired for the full development of the ring capacity (100% in the table). The other limit is the minimum dimension permitted for which some reduction factor is given for the ring capacity. Load capacities for dimensions between these limits can be directly proportioned.

Table 14-2 gives capacities for split-ring connections for both dense and regular grades of Douglas fir–larch and Southern pine. As with bolts, values are given for load directions both parallel to and perpendicular to the grain of the wood. Values for loadings at some angle to the grain can be determined with the use of the Hankinson graph shown in Fig. 13-6.

The following example illustrates the procedures for the analysis of a joint using split-ring connectors.

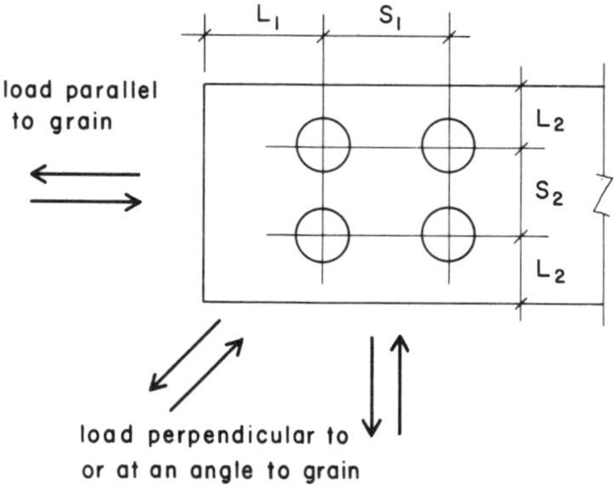

FIGURE 14-4. Reference figure for the end, edge, and spacing requirements for split-ring connectors. (See Table 14-1.)

TABLE 14-1 Spacing, Edge Distance, and End Distance for Split-Ring Connectors

Load Direction with Respect to Grain		Distances (in inches) and Corresponding Percentages of Design Values from Table 14-2: Parallel		Perpendicular or Angle	
Ring size (in.)		2.5	4	2.5	4
L_1	Tension	5.50 in., 100% 2.75 in. min, 62.5%	7 in., 100% 3.50 in. min, 62.5%	5.50 in., 100%	7 in., 100%
	Compression	4 in., 100% 2.50 in. min, 62.5%	5.50 in., 100% 3.25 in. min, 62.5%	2.75 in. min, 62.5%	3.25 in. min, 62.5%
L_2	Unloaded	1.75 in. min	2.75 in. min	1.75 in. min	2.75 in. min
	Loaded[a]				
S_1		3.50 in. min, 50% 6.75 in., 100%	5 in. min, 50% 9 in., 100%	3.50 in. min	5 in. min
S_2		3.50 in. min	5 in. min	3.5 in. min, 50% 4.25 in., 100%	5 in. min, 50% 6 in., 100%

Note: See Fig. 14-4.
[a] See Table 14-2 and Fig. 14-3.

TABLE 14-2 Design Values for Split-Ring Connectors

Ring Size (in.)	Bolt Diameter (in.)	Faces with Connectors[a]	Actual Thickness of Piece (in.)	Load Parallel to Grain Design Value/Connector (lb) Group A Woods[b]	Load Parallel to Grain Design Value/Connector (lb) Group B Woods[b]	Distance to Loaded Edge[c] (in.)	Load Perpendicular to Grain Design Value/Connector (lb) Group A Woods[b]	Load Perpendicular to Grain Design Value/Connector (lb) Group B Woods[b]
2.5	$\frac{1}{2}$	1	1 min	2630	2270	1.75 min	1580	1350
						2.75 or more	1900	1620
			1.5 or more	3160	2730	1.75 min	1900	1620
						2.75 or more	2280	1940
		2	1.5 min	2430	2100	1.75 min	1460	1250
						2.75 or more	1750	1500
			2 or more	3160	2730	1.75 min	1900	1620
						2.75 or more	2280	1940
4	$\frac{3}{4}$	1	1 min	4090	3510	2.75 min	2370	2030
						3.75 or more	2840	2440
			1.5 or more	6020	5160	2.75 min	3490	2990
						3.75 or more	4180	3590
		2	1.5 min	4110	3520	2.75 min	2480	2040
						3.75 or more	2980	2450
			2	4950	4250	2.75 min	2870	2470
						3.75 or more	3440	2960
			2.5	5830	5000	2.75 min	3380	2900
						3.75 or more	4050	3480
			3 or more	6140	5260	2.75 min	3560	3050
						3.75 or more	4270	3660

[a] See Fig. 14-2.

[b] Group A includes Dense grades and Group B regular grades of Douglas fir–larch and Southern pine.

Example. The joint shown in Fig. 14-5, using $2\frac{1}{2}$ in. split-rings and wood of Douglas fir–larch, Dense No. 1 grade, sustains the load indicated. Find the limiting value for the load.

Solution: Separate investigations must be made for the members in this joint. For the 2 × 6,

Load is parallel to the grain.

Rings are in two faces.

Critical dimensions are member thickness of 1.5 in. [38 mm] and end distance of 4 in.

From Table 14-1 we determine that the end distance required for use of the full capacity of the rings is 5.5 in. and that, if the minimum distance of 2.75 in. is used, the capacity must be reduced to 62.5% of the full value. The value to be used for the 4 in.

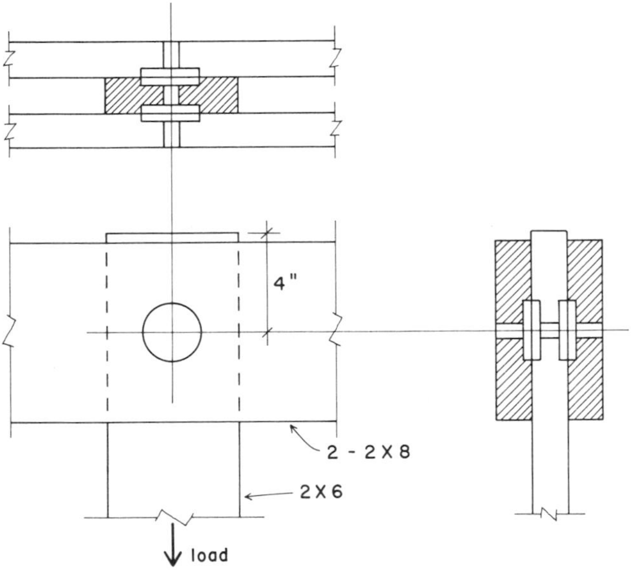

FIGURE 14-5.

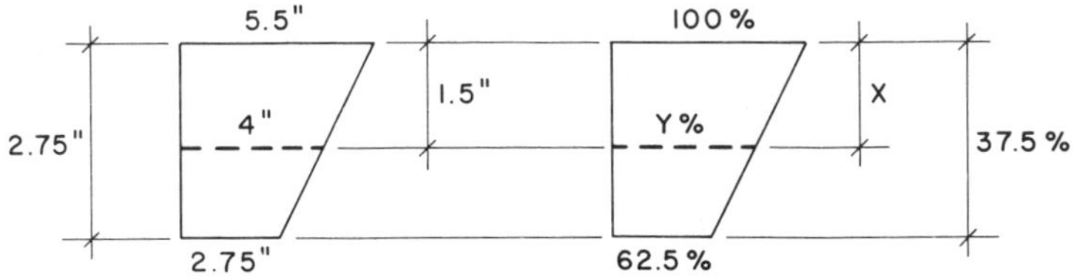

FIGURE 14-6.

end distance must be interpolated between these limits, as shown in Fig. 14-6. Thus

$$\frac{1.5}{x} = \frac{2.75}{37.5} \qquad x = \frac{1.5}{2.75}(37.5) = 20.45\%$$

$$y = 100 - 20.45 = 79.55\%, \text{ or approximately } 80\%$$

From Table 14-2 we determine the full capacity to be 2430 lb per ring. Therefore the usable capacity is

$$(0.80)(2430) = 1944 \text{ lb/ring [8.65 kN/ring]}$$

For the 2 × 8,

Load is perpendicular to the grain.
Rings are in only one face.
Loaded edge distance is one-half of 7.25 in., or 3.625 in. [92 mm].

For this situation the load value from Table 14-2 is 2280 lb [12.81 kN]. Therefore the joint is limited by the conditions for the 2 × 6, and the capacity of the joint with the two rings is

$$T = (2)(1944) = 3888 \text{ lb [17.3 kN]}$$

It should be verified that the 2 × 6 is capable of sustaining this load in tension stress on the net section at the joint. As shown in Fig. 14-7, the net area is

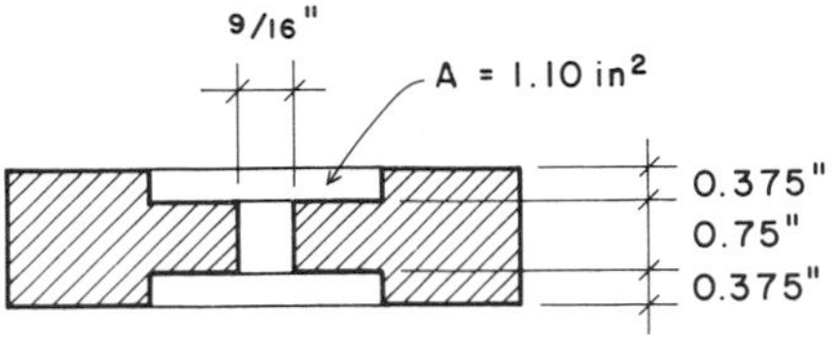

FIGURE 14-7. Determination of the net cross-sectional area.

$$A = 8.25 - (2)(1.10) - \left(\frac{9}{16}\right)(0.75) = 5.63 \text{ in.}^2 \ [3632 \text{ mm}^2]$$

From Table 3-1 the allowable tension stress is 1200 psi [8.27 MPa], and therefore the capacity of the 2 × 6 is

$$T = (1200)(5.63) = 6756 \text{ lb } [30.0 \text{ kN}]$$

and the member is not critical in tension stress.

Problem 14-2-A. A joint similar to that shown in Fig. 14-5 is made with 4 in. split-rings and wood members of Southern pine, No. 1 grade. Find the limit for the tension load if the outer members are each 3 × 10 and the middle member is a 4 × 10.

14-3 Formed Steel Framing Elements

Formed metal-framing devices have been used for many centuries for the assembly of structures of heavy timber. In ancient times elements were formed of bronze or of cast or wrought iron. Later they were formed of forged or bent and welded steel. (See Fig. 14-8.) Some of the devices used today are essentially the same in function and detail to those used long ago. For large timber members, the connections are generally formed of steel plate that is bent and welded to produce the desired form. (See Fig. 14-9.) The ordinary tasks of attaching beams to columns and columns to footings continue to be required and the simple means of achieving the tasks evolve from practical concerns.

For resistance to gravity loads, connections such as those shown in Fig. 14-9 often have no direct structural function. In theory it is possible to simply rest a beam on top of a column as it

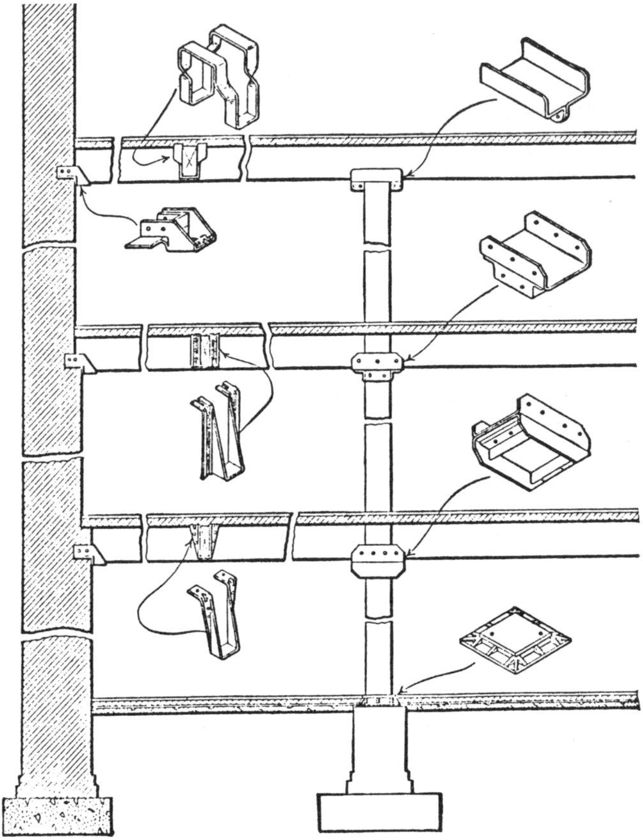

FIGURE 14-8. Formed steel-connecting devices in timber construction. (Reproduced from *Kidder-Parker Handbook,* 18th edition, 1931, with permission of the publishers, John Wiley.)

is done in some rustic construction. However, for resistance to lateral loads from wind or earthquakes, the tying and anchoring functions of these connecting devices are often quite essential. They also serve a practical function in simply holding the structure in position during the construction process.

A development of more recent times is the extension of the use of metal devices for the assembly of light wood frame construc-

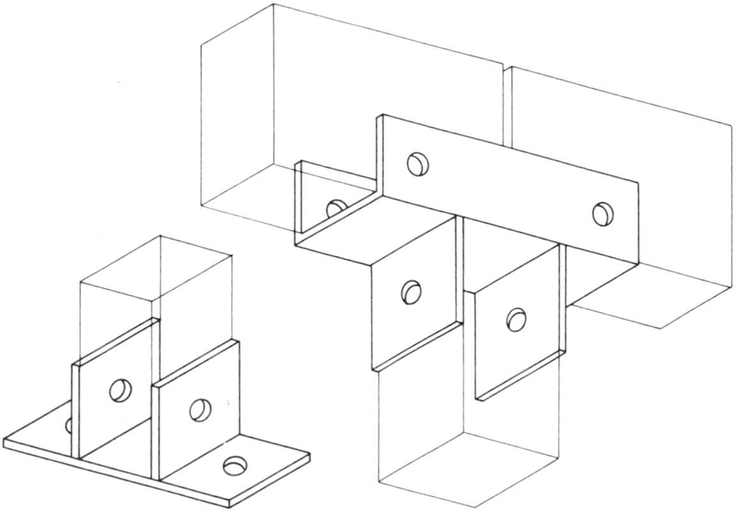

FIGURE 14-9. Formed steel-connecting devices: welded steel plate.

tion. Devices of thin sheet metal, such as those shown in Fig. 14-10, are now quite commonly used for stud and joist construction employing predominantly wood members of 2 in. nominal thickness. As with the devices used for heavy timber construction, these lighter connectors often serve useful functions of tying and anchoring the structure. Load transfers between basic elements of a building's lateral bracing system are often achieved with these elements. (See discussion in Chapter 19.)

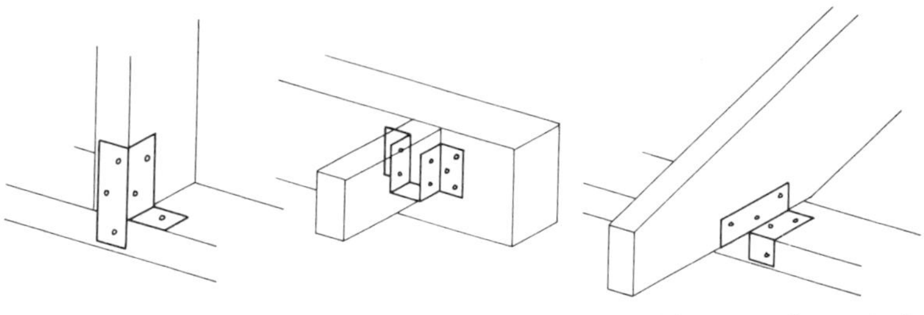

FIGURE 14-10. Formed steel-connecting devices: light gage sheet steel.

Commonly used connection devices of both the light sheet steel type and the heavier steel plate type are readily available from building materials suppliers. Many of these devices are approved by local building codes and have quantified load ratings which are available from the manufacturers. If these ratings are approved by the building code with jurisdiction for a particular building, the devices can be used for computed structural load transfers. Information should be obtained from local suppliers for these items.

For special situations it may be necessary to design a custom-formed framing device. Such devices can often be quickly and easily made up by local metal fabricating shops. However, the catalogs of manufacturers of these devices are filled with a considerable variety of products for all kinds of situations, and it is wise to first determine that the required device is not available as a standard hardware item before proceeding to design it as a special, custom-made item—which is sure to be more expensive in most cases.

14-4 Concrete and Masonry Anchors

Wood members supported by concrete or masonry structures must usually be anchored directly or through some intermediate device. The most common attachment is with steel bolts cast into the concrete or masonry; however, there is also a wide variety of cast-in, drilled-in, or shot-in devices that may be used for various situations. The latter types of anchoring elements are manufactured products, and any data for load capacities or required installation details must be obtained from the manufacturer or supplier.

Two common situations are those shown in Fig. 14-11. The sill member for a wood stud wall is typically attached to a supporting concrete base through steel anchor bolts cast into the concrete. These bolts serve essentially simply to hold the wall securely in position during the construction process. However, they may also serve to anchor the wall against lateral or uplift forces, as is discussed in Chapter 19. For lateral force, the load limit is typically based on the bolt–to–wood limit, as discussed in Sec. 13-2.

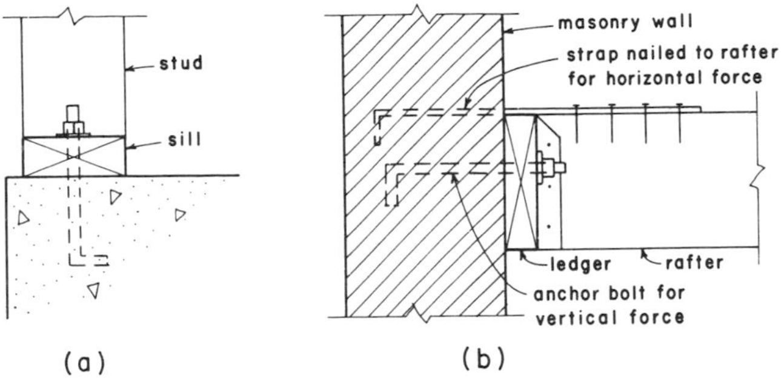

FIGURE 14-11. Anchoring devices for concrete and masonry.

For uplift the limit may be based on tension stress in the bolt or the pullout limit for the bolt in the encasing concrete or masonry.

Figure 14-11*b* shows a common situation in which a wood-framed roof or floor is attached to a masonry wall through a member bolted flat to the wall surface, called a ledger. For vertical load transfer the shear effect on the bolt is essentially as described in Sec. 13-2. For lateral force the problem is one of pullout or tension stress in the bolt, although a critical concern is for the cross-grain bending effect in the ledger. In zones of high seismic risk, it is usually required to have a separate horizontal anchor to avoid the cross-grain bending in the ledger, although the anchor bolts can still be used for gravity loads.

14-5 Plywood Gussets

Cut pieces of plywood are sometimes used as connecting devices, although the increasing variety of formed products for all conceivable purposes has reduced such usage. Trusses consisting of single wood members of a constant thickness are sometimes assembled with gussets of plywood with joints as shown in Fig. 13-9. Although such connections may have considerable load resistance, it is best to be conservative in using them for computed structural forces, especially with regard to tension stress in the plywood.

15

Trusses

Wood trusses are widely used; they are used mostly for roof structures owing to the lighter loads, the less frequent concern for fire resistance, and the ability of the truss to accommodate many forms. This chapter discusses some of the general issues regarding the use of wood trusses. The complete design of a truss system and the design of all the members and joints for a single truss are quite laborious and therefore space limitations here do not permit including a design example. Readers wishing a more extensive treatment of this topic are referred to *Simplified Design of Building Trusses for Architects and Builders* (Ref. 10).

15-1 General

A *truss* is a framed structure with a system of members so arranged and secured to one another that the stresses transmitted from one member to another are either axial compression or tension. Basically a truss is composed of a system of triangles because a triangle is the only polygon whose shape cannot be changed without changing the length of one or more of its sides.

With respect to roofs supported by trusses a bay is that portion of the roof structure bounded by two adjacent trusses; the spacing of the trusses on centers is the width of the bay. A *purlin* is a beam spanning from truss to truss that transmits to the trusses the loads due to snow, wind, and weight of the roof construction. The portion of a truss that occurs between two adjacent joints of the upper chord is called a *panel*. The load brought to an upper-chord joint or *panel point* is therefore the roof design load in pounds per square foot multiplied by the panel length times the bay width; this is called a *panel load*. Figure 15-1*a* gives the names of other elements of a typical roof truss.

15-2 Types of Trusses

Figure 15-2 shows some of the more common roof trusses. The height or *rise* of a truss divided by the span is called the *pitch;* the rise divided by half the span is the *slope*. Unfortunately these two terms are often used interchangeably. A less ambiguous way of expressing the slope is to give the amount of rise per foot of span. A roof that rises 6 in. in a horizontal distance of 12 in. has a slope of "6 in 12." Reference to Table 15-1 should clarify this terminology.

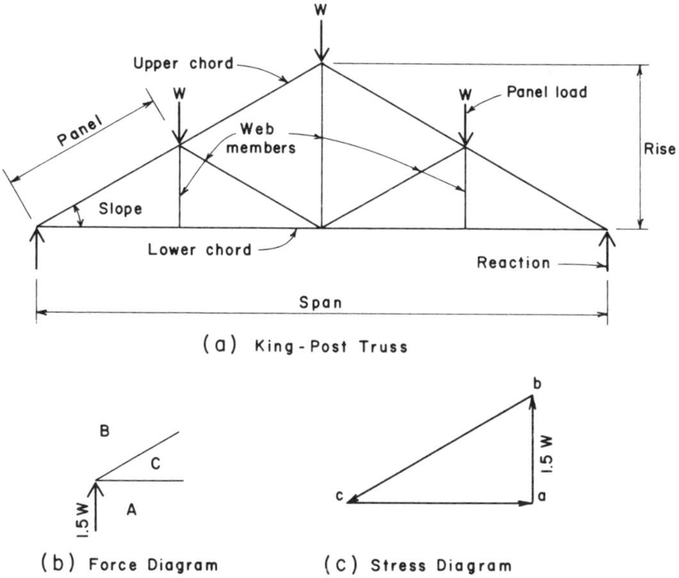

FIGURE 15-1.

TABLE 15-1. Roof Pitches and Slopes

Pitch	$\frac{1}{8}$	$\frac{1}{6}$	$\frac{1}{5}$	$\frac{1}{4}$	1/3.46	$\frac{1}{3}$	$\frac{1}{2}$
Degrees	14°3′	18°26′	21°48′	26°34′	30°0′	33°40′	45°0′
Slope	3 in 12	4 in 12	4.8 in 12	6 in 12	6.92 in 12	8 in 12	12 in 12

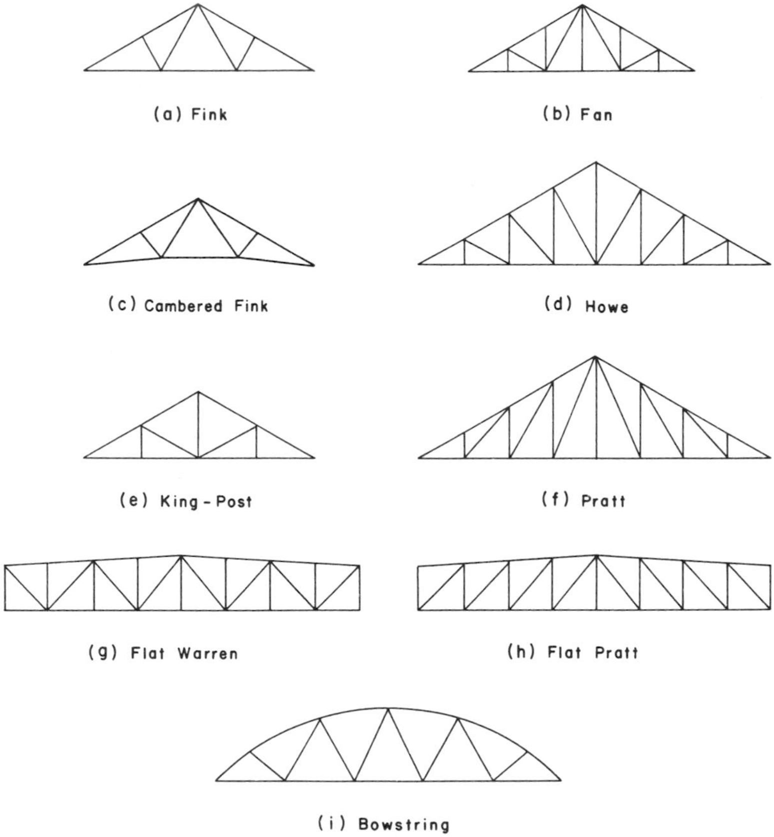

FIGURE 15-2. Types of wood trusses.

15-3 Stresses in Truss Members

When designing a roof truss, the designer must first determine the magnitude and character of the stress in each member. By character is meant the kind of stress–tension or compression. This may be accomplished by graphical methods.

With the exception of wind loads, trusses are generally symmetrically loaded. In the truss shown in Fig. 15-1*a*, for example, there would be three equal vertical panel loads *W* lb each; and, because the truss is symmetrical, each upward force or reaction at the supports would equal $\frac{1}{2} \times 3W = 1\frac{1}{2}W$. The lower left-hand end, or heel joint, of the truss is represented diagrammatically in Fig. 15-1*b*. The forces, read clockwise about the joint, are three in number: *AB* is the reaction, an upward force of $1\frac{1}{2}W$ lb; *BC* the upper chord and *CA* the lower chord, both of which are of unknown magnitude. These forces are concurrent, and because, by data, they are in equilibrium a stress diagram corresponding to the forces must close. Therefore draw an upward force *ab* to some convenient scale representing the reaction $1\frac{1}{2}W$ (Fig. 15-1*c*). From point *b* draw a line parallel to *BC* and from point *a* draw a line parallel to *CA*. The intersection of these two lines determines point *c*. The magnitudes of the stresses in the members *BC* and *CA* are found by scaling their lengths in the stress diagram (Fig. 15-1*c*) at the same scale at which *ab* was drawn.

To determine the character of the stresses at this joint we refer to Figs. 15-1*b* and *c*. Consider first the member *BC* about the point *ABC*. Because the forces were read in a clockwise direction, note the sequence of the letters: *B* first, then *C*. In the stress diagram we find that *bc* reads downward to the left. In the force diagram, if *BC* reads downward to the left, this is *toward* the reference point *ABC;* hence the member *BC* is in compression. Next consider the member *CA* about the point *ABC*. In the stress diagram we find that *ca* reads from left to right. If we consider the member *CA* in the force diagram as reading from left to right, we read *away* from the reference point *ABC;* therefore the member *CA* is in tension.

The length of a member in a truss bears no direct relation to the magnitude of its stress. The magnitude of the stress is determined by the length of the line in the stress diagram corresponding to the truss member.

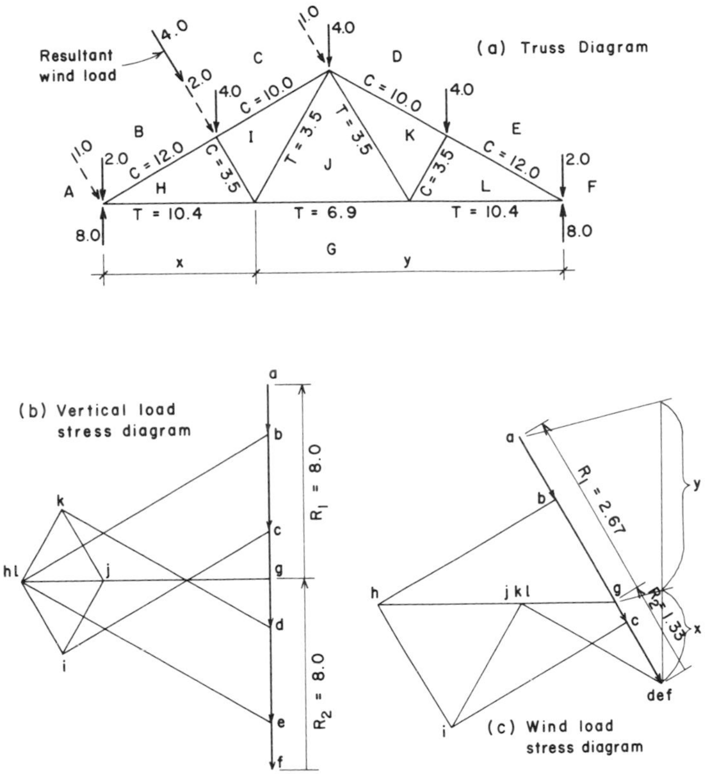

FIGURE 15-3. Truss and stress diagrams (loads and member stresses in kips).

The stress diagram (Fig. 15-1*c*) is the force polygon for the three concurrent forces at the heel of the truss. The stress diagram for the entire truss would consist of combined force polygons for the forces at all the different joints of the truss. Figure 15-3*a* is a four-panel Fink truss; the complete stress diagram for vertical loads is shown in Fig. 15-3*b*.

15-4 Stress Diagrams

The panel load for the vertical loading on the truss shown in Fig. 15-3*a* is 4 kips. The two end loads are 2 kips each, making a total vertical load of 4 + 4 + 4 + 2 + 2 = 16 kips. Because the truss is symmetrical, each upward force at the support (reaction) is 16 ÷ 2 = 8 kips.

The panel loads and reactions being known, *the first step in constructing a stress diagram is to draw a force polygon of the external forces*. These forces are *AB, BC, CD, DE, EF, FG,* and *GA,* and the magnitudes of all are known. Therefore, at a suitable scale, draw *ab* (Fig. 15-3*b*), a downward force equal to 2 kips. The next external force is *BC,* and from point *b,* just determined, draw *bc* downward, equivalent to 4 kips. Continue with *CD, DE,* and *EF*. This completes the downward forces. The line just drawn is called the *load line,* the length being equivalent to 16 kips. The next external force is *FG,* an upward force of 8 kips. This determines the position of point *g;* and *GA,* the remaining external force, completes the polygon of the external forces. Because the loads and reactions are vertical, the force polygon of the external forces is a vertical line.

In conjunction with the polygon just drawn, draw a force polygon for the forces *AB, BH, HG,* and *GA* about the point *ABHG.* From *b* draw a line parallel to *BH;* and from *g* draw a line parallel to *HG*. Their intersection determines point *h*. Next consider the members about the joint *BCIH*. From *c* draw a line parallel to *CI;* and through *h* draw a line parallel to *IH;* their intersection determines point *i*. The next joint is *HIJG*. Through *i* draw a line parallel to *IJ* and through *g* draw a line parallel to *JG,* Thus establishing point *j*. In the same manner take the joints *CDKJI* and *DELK*. This completes the stress diagram. (See Fig. 15-3*b*.) The magnitudes of the stresses in the members are found by scaling the lengths of the lines in the stress diagram just completed. The character of the stresses is found as explained in Sec. 15-3. When recording the character of the stresses, a minus sign (−) usually denotes compression and a plus sign (+) denotes tension unless otherwise noted. In some books, however, these designations are reversed. To avoid confusion the system employed in this text uses the symbol *C* for compressive stress and *T* for tensile stress. The important point to remember is that a member in compression tends to be made shorter and resists this shortening by *pushing against* the joints at its ends. A tension member, on the other hand, tends to become longer and resists the lengthening by *pulling away* from its end joints. For the vertical loads the character and magnitude of the stresses are shown on the truss diagram of Fig. 15-3*a*.

Wind Load Stress Diagram. The stresses in the members of a roof truss are found by constructing a stress diagram as previously described. It is assumed that the wind exerts a pressure perpendicular to the roof surface. In the truss diagram (Fig. 15-3*a*) the wind, indicated by dotted lines, is shown coming from the left, with the total load 1 + 2 + 1 = 4 kips. To draw a stress diagram for the wind loads proceed as before. Construct the force polygon of the external forces, namely, *AB, BC, C–DEF, DEF–G,* and *GA;* the latter two are wind-load reactions. Note that because the wind comes from the left there are no forces *DE* and *EF;* consequently the letters *D, E,* and *F* represent a single point in the stress diagram. Draw *ab, bc,* and *c–def.* (See Fig. 15-3*c*.) It may be assumed that the reactions due to wind loads are parallel to the direction of the wind, and because the wind comes from the left, the left-hand reaction will have a greater magnitude than that on the right for this particular truss. For the purpose of finding the magnitudes of the reactions we may consider that all the wind loads are concentrated in a single line of action at *BC*. The resultant wind load continued divides the lower chord into two parts, *x* and *y*, and the magnitudes of the reactions are to each other as the division *x* is to *y*. The line *a–def,* representing the total wind load, is therefore divided in the same proportion as the two divisions of the lower chord *x* and *y*. To accomplish this division erect a line from the point *def* in a length equivalent to the length of the lower chord and divide it into sections *x* and *y*. From the upper extremity of the line just drawn add a line to point *a* and a parallel line from the point separating *x* and *y* to the load line. This determines point *g* and consequently the reactions R_1 and R_2. The force polygon of the external forces is now completed, and the stress diagram is then drawn as previously described. It will be found that the letters *j, k,* and *l* fall at one point, indicating that there are no stresses in *JK* and *KL* when the wind comes from the left. For design purposes, however, the stresses in *JK* and *KL* are taken the same as for *JI* and *IH*, respectively, because the direction of the wind may be reversed.

15-5 Truss Members and Joints

The three common forms of truss configuration are those shown in Fig. 15-4. The single member type, with all members in one plane as shown in Fig. 15-4*a*, is that used most often to produce

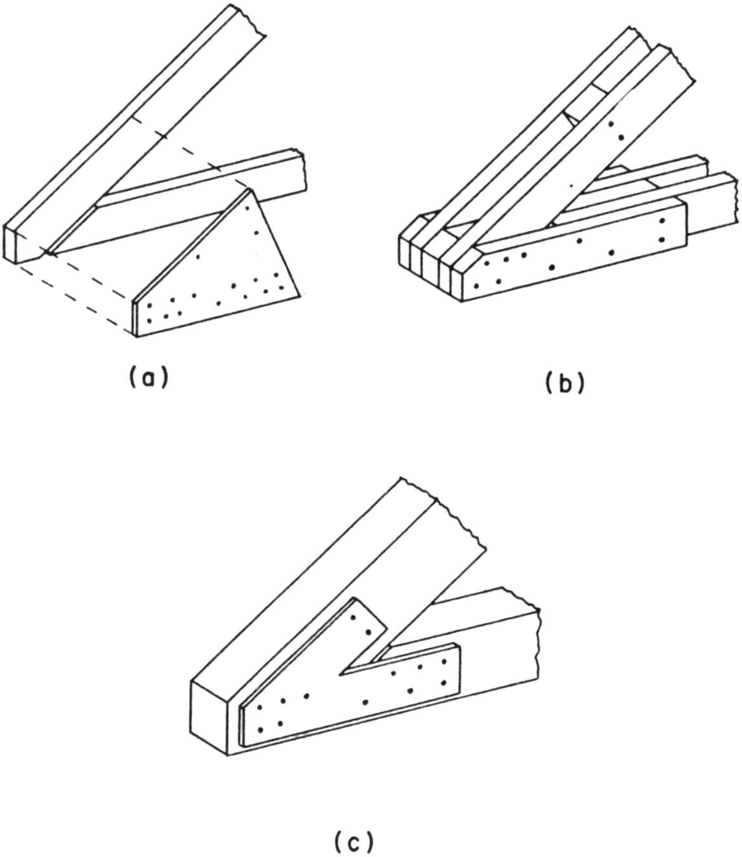

FIGURE 15-4. Typical forms of wood trusses.

the simple W-form truss (Fig. 15-2*a*), with members usually of 2 in. nominal thickness. Joints may use plywood gussets as shown in Fig. 15-4*a*, but are more often made with metal-connecting devices when the trusses are produced as standard products by a manufacturer. In the latter case the joint performance is certified by load testing of prototypes.

For larger trusses the form shown in Fig. 15-4*b* may be used, with members consisting of multiples of standard lumber elements. If the member carries compression, it will usually be designed as a spaced column, as described in Sec. 12-8. For modest spans members are usually of two elements of nominal 2 in. thickness. However, for large spans or heavy loads, individual members may consist of several pieces with thicknesses greater than 2 in. nominal. Joints are usually made with bolts and some form of shear developer such as split rings.

In the so-called *heavy timber truss* the individual members are of large timber elements, usually occurring in a single plane as shown in Fig. 15-4*c*. A common type of joint in this case is one using steel plates attached by lag screws or through-bolts. Depending on the truss pattern and loading, it may be possible to make some joints without the gussets, as is shown in Fig. 15-5*a* for the diagonal compression member and the bottom chord con-

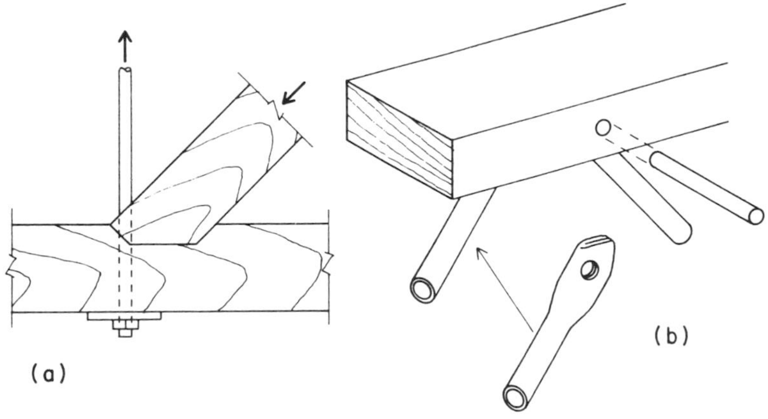

FIGURE 15-5. Forms of composite wood and steel trusses.

nection. This was common in the past, but it requires carpentry work that is not easily obtained today.

Although wood members have considerable capability to resist tension, the achieving of tension connections is not so easy, especially in heavy timber trusses. Thus in some trusses the tension members are made of steel, as shown for the vertical member in Fig. 15-5*a*. A common truss form today is that shown in Fig. 15-5*b* where a flat-chorded truss has chords of wood and all interior members of steel.

The selection of truss members and jointing methods depends on the size of the truss and the loading conditions. Unless trusses are exposed to view and appearance is a major concern, the specific choices of members and details of the fabrication are often left to the discretion of the fabricators of the trusses.

15-6 Heavy Timber Trusses

Single members of large dimension timber were used for trusses many years ago. As the iron and steel technology developed, cast iron and steel elements were employed for various tasks in this type of construction. Figure 15-6 shows an example of this type of structure as it was developed up to the early part of this century. This type of construction—and particularly the jointing methods—is seldom used today except in restoration or reproduction of historic buildings. The special parts, such as the cast iron shoe and the highly crafted notched joints, are generally unattainable.

Trusses of the size and general form of those shown in Fig. 15-6 are now mostly made with multiple-element members and bolted joints, as shown in Fig. 15-5*b*. Although steel elements may be used for some tension members, this is done less frequently today because joints using shear developers are capable of sufficient load resistance to permit use of wood tension members. It was chiefly the connection problem that inspired the use of steel rods in the timber truss. The one vestige of this composite construction today is the use of steel members in the manufactured trusses of the form shown in Fig. 15-5*b*.

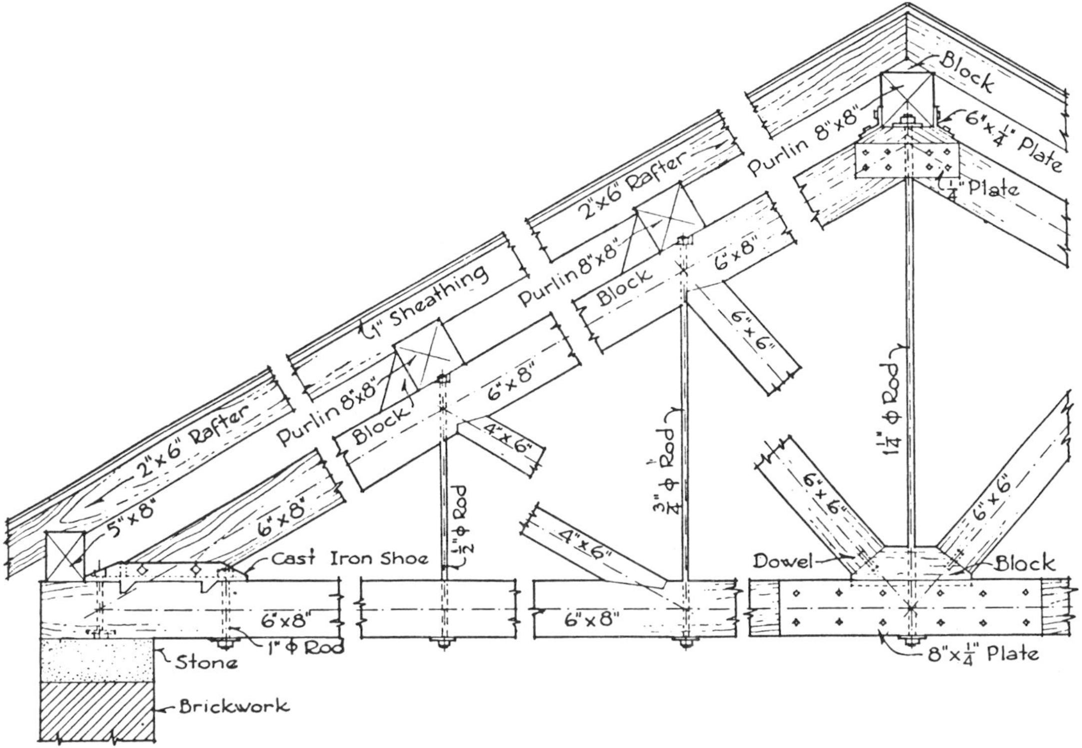

FIGURE 15-6. Typical form of old heavy timber truss. (From *Materials and Methods of Architectural Construction* by Gay and Parker, 1932, reproduced with permission of the publishers, John Wiley.)

15-7 Manufactured Trusses

The majority of trusses used for building structures in the United States today are produced as manufactured products, and the detailed engineering design is done largely by the engineers in the employ of the manufacturers. Since shipping of large trusses over great distances is not generally feasible, the use of these products is chiefly limited to some region within a reasonable distance from a particular manufacturer. If such a product is to be used, it must first be established as to which particular products are available in the region of the building site.

There are three principal types of manufactured truss. Manufacturers vary in size of operation; some produce only one type and others have a range of products. These three types are as follows:

1. One type is the simple gable-form, W-pattern truss (Fig. 15-2*a*) with truss members of single-piece, 2 in. nominal lumber. These are quite simple to produce and can be turned out in small quantities by neighborhood building suppliers in some instances. Larger companies use automated processes and more sophisticated handling in general, but the range of availability is generally extended due to the simplicity of the product.
2. Second is the so-called trussed joist, usually consisting of some type of composite wood and steel elements, as shown in Fig. 15-5*b*. These are more sophisticated in fabrication detail than the simple W-trusses, and they are mostly produced by larger companies. They compete generally with steel open web joists, with a competitive margin that varies from region–to–region.
3. The third type is the large, long-span truss, usually using multiple-element members, as shown in Fig. 15-4*b*. These may be produced with some standardization by a specific manufacturer, but are typically customized to some degree for a particular building. One form used for very long spans is the bowstring type, which in reality is essentially a tied arch. (See Fig. 15-2*i*.)

Suppliers of manufactured trusses usually have some fairly standardized models, but almost always have some degree of variability to accommodate the specific usage considerations for a particular building. The possible range of these variations can only be established by working with the suppliers.

16

Glued-Laminated Products

The basic process of gluing together pieces of wood that is used to produce plywood panels can be extended to other applications. This chapter deals with some of the other products that are produced by glue laminating for structural applications in buildings.

16-1 Types and Usages of Products

For structural applications the principal types of products and their typical uses are the following (see Fig. 16-1):

1. *Multiple Laminated Structural Lumber.* These elements are produced by laminating multiples of standard 2× or 1× members. The most widely used product is the beam or girder that is produced from multiple 2× pieces in a form similar to solid-sawn timbers. However, laminated elements may also be curved and the 1× lumber is used where a shorter radius of curvature is desired. The size of cross sections and the lengths are virtually unlimited and many huge elements have been produced by this process.

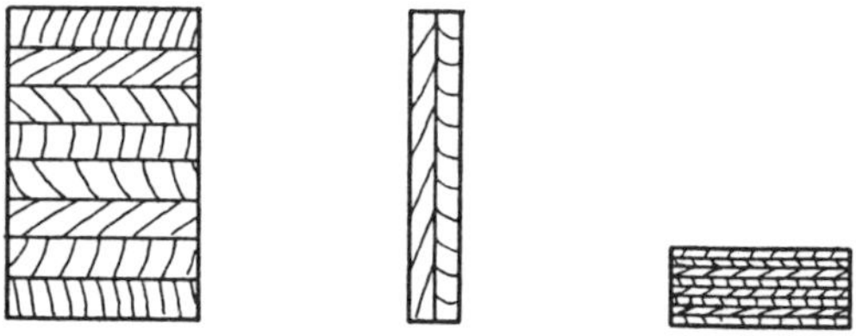

(a) (b) (c)

FIGURE 16-1. Forms of glued-laminated structural elements.

2. *Vertically Laminated Joists.* These usually consist of $\frac{3}{4}$ in. thick (nominal 1×) boards laminated in multiples of two or three for use as joists to compete with the high end of the size range of standard 2× and 3× lumber. (See Fig. 16-1*b*.)
3. *Microlaminated Elements.* These consist of small size cross sections produced with laminations of about 0.1 in. thickness. A major structural use is for the chords of manufactured trusses where high strength, dimensional stability, and virtually unlimited, single-piece length are all important considerations. (See Fig. 16-1*c*.)

16-2 Beams and Girders

Lamination of standard 2 in. nominal thickness lumber has been used for many years to produce large beams and girders. This is really the only option for a single-piece wood member beyond the feasible range of size for solid-sawn lumber. However, there are other reasons for using the laminated beam that include the following:

1. *Higher Strength.* Lumber used for laminating consists of a moisture content described as kiln dried. This is the opposite end of the quality range from the green wood condition assumed for large solid-sawn members. This, plus the minimizing effect of flaws due to lamination, permits use of stresses for flexure and shear that are as much as twice those allowed for ordinary grades of solid-sawn lumber.

The result is that much smaller sections can often be used, which may be enough to justify the relatively high cost of the laminated products.

2. *Better Dimensional Stability.* This refers to the tendency to warp, split, shrink, and so on. Both the use of the kiln-dried materials and the laminating process itself tend to create a very stable product. This is often a major consideration where shape change can adversely affect the building construction.
3. *Shape Variability.* Lamination permits the production of curved, tapered, and other profile forms for a beam, as shown in Fig. 16-2. Cambering for dead load deflections, sloping for roof drainage, and other desired custom profiling can be done with relative ease. This is otherwise possible only with a truss or a built-up section.

Laminated beams have seen wide use for many years and industrywide standards are well established. (See the *Timber Construction Manual,* Ref. 2, or Chapter 25 of the *Uniform Building Code,* Ref. 3.) Sizes are generally established by the process and the size of standard lumber elements used for lamination. Thus depths of sections are multiples of 1.5 in. and widths are slightly less than the lumber size as a result of the finishing of the product. Minor misalignment of laminations and the sloppiness of the gluing result in a generally unattractive surface. Finishing may simply consist of a smoothing off, although special finishes, such as rough-sawn ones, are also available.

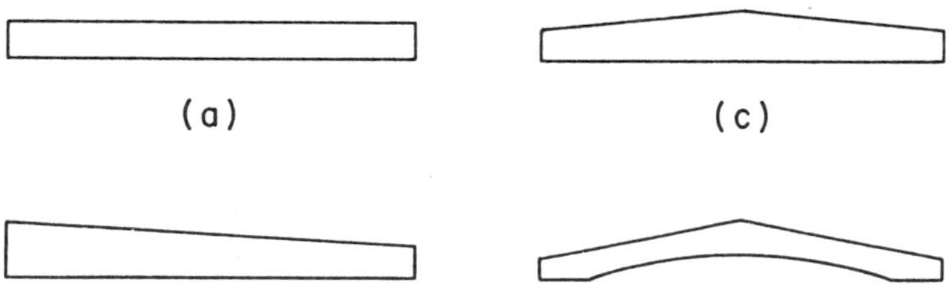

FIGURE 16-2. Profile variations for glued laminated timbers.

These elements are manufactured products and information about them should be obtained from suppliers that serve the area in the region of the building site. There is considerable industry standardization and extensive building code control, but it is still best to know the particular products actually available in a given location.

16-3 Arches and Bents

Individual elements of glued-laminated lumber can be custom-profiled to produce a wide variety of shapes for structures. Two forms commonly used are the three-hinged arch (Fig. 16-3*a*) and the gabled bent (Fig. 16-3*b*). A critical consideration is that of the radius of curvature of the member, which must be limited to what the wood species and the laminate thickness can tolerate. For very large elements this is not a problem, but for smaller structures the bending limits of 2 in. nominal lumber may be critical.

Manufacturers of laminated products usually produce the arch and gabled bent elements as more or less standard items. Actual structural design of the products is most often done by the manufacturer's engineers. Form limits, size ranges, connection details, and other considerations affecting design should be investigated with individual manufacturers.

Other shapes, of course, are possible, such as that of the doubly curved elements shown in Fig. 16-3*c*. Many imaginative structures have been designed using the form variation potential of this process.

16-4 Laminated Columns

Columns consisting of multiples of 2 in. nominal lumber are sometimes used. The advantages of the higher strength material may be significant, especially if combined resistance to bending and axial compression must be developed. In some situations the increased dimensional stability may be a major advantage.

The large glued-laminated column section offers the same general advantages of fire resistance that are given to heavy timber construction with solid-sawn sections. This does not make it

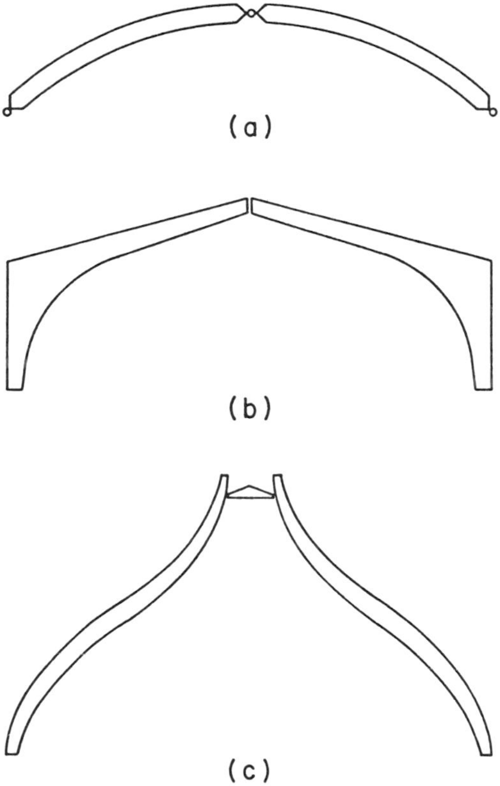

FIGURE 16-3. Nonstraight laminated elements for spanning structures: (*a*) arch segments; (*b*) gabled frame; (*c*) custom formed, doubly curved elements.

more competitive in comparison to the solid wood section but does in comparison to an exposed steel column.

It is also possible to produce columns of great single-piece length, of tapered form, of considerable width, and so on. In other words, the advantages include all the potential offered for other laminated structural elements. In general, laminated columns are used less frequently than beams or girders, and are mostly chosen only when some of the inherent limitations of other options for the column are restrictive.

A special application of the laminated column is in built-up sections, such as that shown in Fig. 12-7*b*. In these situations the laminated portion is considered as the functioning structural element, and the added solid-sawn elements are limited to decorative functions or use for other construction reasons.

17

Plywood

17-1 General

Plywood is the term used to designate wood panels made by gluing together multiple layers of thin wood veneer (called *plies*) with alternate layers having their grain direction at right angles. The outside layers are called the *faces* and the others, *inner plies*. Inner plies with the grain direction perpendicular to the faces are called *crossbands*. There is usually an odd number of plies so that the faces have the grain in the same direction. For structural uses as wall sheathing or roof or floor decks, the common range of panel thickness is $\frac{1}{4}$–$1\frac{1}{8}$ in. and the usual panel size is 4 ft by 8 ft.

The alternating grain direction of the plies gives the panels considerable resistance to splitting, and as the number of plies is increased the panels become approximately equally strong in both directions. Thinner panels are most effective when spanning with the grain direction of the face plies perpendicular to the supports. For $\frac{3}{4}$ in. and greater thicknesses this becomes less critical.

17-2 Types and Grades of Plywood

Structural plywood consists primarily of that made with all plies of Douglas fir. Many different kinds of panels are produced; the principal distinctions other than the panel thickness are the following:

1. *Glue Type*. Panels are identified for exterior (exposed to the weather) or interior use based on the type of glue used. Exterior type should also be used for any interior conditions involving high moisture.
2. *Grade of Plies*. Individual plies are rated—generally A, B, C, or D with A best—on the basis of the presence of knots, splits, patches, or plugs. The most common concern is for the quality of the face plies; thus a panel is typically rated on the basis of the front and back plies. For example, a designation of C–C indicates that both faces are of C grade; a designation of C–D indicates the front is C grade and the back is D grade.
3. *Structural Classification*. In some cases panels are identified as Structural I or Structural II. This is mostly of concern when panels are used for shear walls or for horizontal roof or floor diaphragms. For this rating the grades of the inner plies are also considered.
4. *Special Faces*. Plywood with special facing, usually only on one side, is produced for a variety of uses. Special surfaces for use as exposed finish siding are an example. These are usually the products of a particular manufacturer and structural properties and usage considerations should be obtained from the supplier.

Some ratings and classifications are industrywide and some are local variations due to the use of a particular building code or the regional use of particular products. Designers should be aware of general industry standards, but should also be informed about the products that are generally available and frequently used in a given locality.

17-3 Panel Identification Index

Structural grades of plywood usually have a designation called the *Identification Index,* which is stamped on the panel back as part of the grade trademark. This index is a measure of the strength and stiffness of the panel and consists of two numbers separated by a slash (/). The first number indicates the maximum center–to–center spacing of supports for a roof deck under average loading conditions and the second number indicates the maximum spacing for a floor deck for average residential loading. There are various conditions on the use of these numbers, but they generally permit the selection of panels for a specific ordinary situation without structural computations.

17-4 Usage Data

Data for structural design with plywood may be obtained from industry publications or from individual plywood producers. Some building codes have data for plywood design for ordinary situations, usually developed from industry publications. Table 17-1 is a reprint of Table 25-S-1 from the *Uniform Building Code* (Ref. 3) and includes data for plywood decks where the face grain is perpendicular to the supports. In similar form, Table 17-2 is a reprint of Table 25-S-2 from the *Uniform Building Code* and gives data for decks where the face grain is parallel to the supports. Footnotes to these tables present various qualifications including some of the loading and deflection criteria. There are other panel thicknesses and types of panels not covered by these tables that may be designed from data available from manufacturers or suppliers. Code acceptability of such data should always be assured before use for any design work.

17-5 Plywood Diaphragms

Plywood roof and floor decks are frequently used as horizontal diaphragms as part of the lateral bracing system for a building. Selection of the plywood type, panel thickness, and particularly

TABLE 17-1. Allowable Spans for Plywood Subfloor and Roof Sheathing Continuous Over Two or More Spans and Face Grain Perpendicular to Supports[1,8]

PANEL SPAN RATING[3]	PLYWOOD THICKNESS (Inch)	ROOF[2] Maximum Span (In Inches)		ROOF[2] Load (In Pounds per Square Foot)		FLOOR MAXIMUM SPAN[4] (In Inches)
		Edges Blocked	Edges Unblocked	Total Load	Live Load	
1. 12/0	15/16	12		135	130	0
2. 16/0	5/16, 3/8	16		80	65	0
3. 20/0	5/16, 3/8	20		70	55	0
4. 24/0	3/8	24	16	60	45	0
5. 24/0	15/32, 1/2	24	24	60	45	0
6. 32/16	15/32, 1/2, 19/32, 5/8	32	28	55	35[5]	16[7]
7. 40/20	19/32, 5/8, 23/32, 3/4, 7/8	40	32	40[5]	35[5]	20[7,8]
8. 48/24	23/32, 3/4, 7/8	48	36	40[5]	35[5]	24

[1]These values apply for C-C, C-D, Structural I and II grades only. Spans shall be limited to values shown because of possible effect of concentrated loads.

[2]Uniform load deflection limitations 1/180 of the span under live load plus dead load, 1/240 under live load only. Edges may be blocked with lumber or other approved type of edge support.

[3]Span rating appears on all panels in the construction grades listed in Footnote No. 1.

[4]Plywood edges shall have approved tongue-and-groove joints or shall be supported with blocking unless 1/4-inch minimum thickness underlayment, or 1½ inches of approved cellular or lightweight concrete is placed over the subfloor, or finish floor is 25/32-inch wood strip. Allowable uniform load based on deflection of 1/360 of span is 165 pounds per square foot.

[5]For roof live load of 40 pounds per square foot or total load of 55 pounds per square foot, decrease spans by 13 percent or use panel with next greater span rating.

[6]May be 24 inches if 25/32-inch wood strip flooring is installed at right angles to joists.

[7]May be 24 inches where a minimum of 1½ inches of approved cellular or lightweight concrete is placed over the subfloor and the plywood sheathing is manufactured with exterior glue.

[8]Floor or roof sheathing conforming with this table shall be deemed to meet the design criteria of Section 2516.

Source: Reproduced from the *Uniform Building Code,* 1985 edition (Ref. 3), with permission of the publishers, International Conference of Building Officials.

TABLE 17-2. Allowable Loads for Plywood Roof Sheathing Continuous Over Two or More Spans and Face Grain Parallel to Supports[1,2]

	THICKNESS	NO. OF PLIES	SPAN	TOTAL LOAD	LIVE LOAD
STRUCTURAL I	15/32	4	24	30	20
		5	24	45	35
	1/2	4	24	35	25
		5	24	55	40
Other grades covered in U.B.C. Standard No. 25-9	15/32	5	24	25	20
	1/2	5	24	30	25
	19/32	4	24	35	25
		5	24	50	40
	5/8	4	24	40	30
		5	24	55	45

[1]Uniform load deflection limitations: 1/180 of span under live load plus dead load, 1/240 under live load only. Edges shall be blocked with lumber or other approved type of edge supports.

[2]Roof sheathing conforming with this table shall be deemed to meet the design criteria of Section 2516.

Source: Reproduced from the *Uniform Building Code,* 1985 edition (Ref. 3), with permission of the publishers, International Conference of Building Officials.

the nailing may be strongly determined by this required function. This topic is developed in Chapter 19.

17-6 Usage Considerations

There are many uses for plywood in building construction and a broad array of products available. For ordinary structural applications the following are some of the principal usage considerations.

1. *Choice of Type and Grade.* This is largely a matter of common usage and building code acceptability. For economy the thinnest, lowest grade panels will always be used for structural applications unless other nonstructural concerns

require better face grades or some specific need. Roof decks may be required to be a minimum thickness to accommodate the particular roofing materials.

2. *Modular Supports.* With the usual common panel size of 4 ft by 8 ft, logical spacing of studs, rafters, and joists become full number divisions of the 48 or 96 in. dimensions: 12, 16, 24, 32, or 48. However, spacing of framing must also often relate to what is on the other side of a wall or on a ceiling surface.
3. *Blocking.* Panel edges not falling on a support may need some backup for nailing, especially for roof and floor decks. Blocking is the common answer, although thick deck panels may have tongue–and–groove edges in special applications.
4. *Attachment.* Attachment of structural plywood is still mostly done with ordinary wire nails. Required nail size and spacing relate to panel thickness and code specifications. Where diaphragm action is required, it often determines the nailing requirements. Gluing of panels to supports is done in some cases, but is seldom relied on exclusively. Nailing of large decks is increasingly done with mechanical driving devices, and the driven object is not always an ordinary nail, although load ratings are usually in terms of equivalency to ordinary nails.
5. *Fabricated Units.* Various types of construction components are produced for use as wall, roof, or floor units consisting of assemblages of plywood and solid wood elements. Some of these units are discussed in Sec. 18.1.

18

Special Problems and Miscellaneous Elements

18-1 Notched Beams

It is frequently necessary to notch the ends of beams to increase clearance or to bring top surfaces level with adjacent beams or girders. Sometimes they are notched, top or bottom, at points between the supports to provide room for pipes or for framing of other beams. Notches in beams reduce the cross-sectional areas and reduce the magnitudes of the loads that the beam will properly support. Several types of notch are shown in Fig. 18-1. With the exception of the beam illustrated in Fig. 18-1*a*, the stiffness of a beam is almost unaffected by notches. When the notches are cut at or near the midpoint in the open length of a beam, the net depth should be used in computing the bending strength.

Shear Resistance in Notched Beams. When a beam is notched on the lower side at the ends, as in Fig. 18-1*b*, the strength of a short, relatively deep beam is decreased by an amount that depends on the shape of the notch and on the relation of the depth of the notch to the depth of the beam. The actual

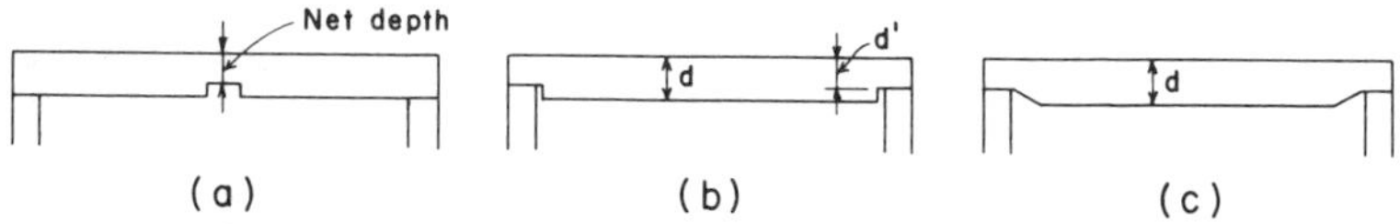

FIGURE 18-1.

depth of material *above the notch* is used as the effective depth in the shear formula. In designing beams with *square-cornered* notches at the ends, the bending load should be checked for shear by the shear formula

$$V = \left(\frac{2F_v bd'}{3}\right) \times \left(\frac{d'}{d}\right)$$

in which V = vertical design shear (lb),
b = width of beam (in.),
d' = depth of beam remaining at notch (in.),
d = total depth of beam (in.),
F_v = design value (allowable unit stress) in horizontal shear (psi).

Example. A 6 × 12 beam has a span of 14 ft with a concentrated load at the center of the span of 4400 lb. The beam has square notches at the supports similar to that shown in Fig. 18-1*b* and the depth of the beam above the notch is $d' = 8$ in. The allowable unit horizontal shearing stress of the timber is 135 psi and the allowable extreme fiber stress is 1600 psi. (1) Is the beam safe with respect to horizontal shear? (2) Is the beam safe with respect to strength in bending?

Solution: (1) To find the maximum permissible vertical shear we use the formula previously stated. Recording the quantities needed, we note that a 6 × 12 timber has dressed dimensions of 5.5 in. × 11.5 in. Then, for the conditions in this problem, $b = 5.5$ in., $d' = 8$ in., $d = 11.5$ in., and $F_v = 135$ psi. Substituting in the formula,

$$V = \left(\frac{2F_v bd'}{3}\right) \times \left(\frac{d'}{d}\right) = \left(\frac{2 \times 135 \times 5.5 \times 8}{3}\right) \times \left(\frac{8}{11.5}\right)$$

$$= 2755 \text{ lb}$$

(2) From Table 4-1 the beam weight is 15.4 lb per lin ft. Then (2 $\times$ V) $-$ (weight of beam) $=$ (2 $\times$ 2755) $-$ (15.4 $\times$ 14) $=$ 5294 lb, the magnitude of the concentrated load that can be placed at the center of the span. Because this notched beam can safely resist a concentrated load of 5294 lb, the 4400 lb load is safe with respect to horizontal shear.

(3) The maximum bending moment for the concentrated load is

$$M = \frac{PL}{4} = \frac{4400 \times 14 \times 12}{4} = 184{,}800 \text{ in.-lb}$$

and that for the distributed load is

$$M = \frac{WL}{8} = \frac{(15.4 \times 14) \times 14 \times 12}{8} = 4528 \text{ in.-lb}$$

which makes the maximum bending moment 184,800 $+$ 4528 $=$ 189,328 in.-lb.

(4) The section modulus of a 6 $\times$ 12 beam is 121.2 in.^{3} (Table 4-1). Therefore the value of the extreme fiber stress developed by the loading is

$$f_b = \frac{M}{S} = \frac{189{,}328}{121.2} = 1562 \text{ psi}$$

Because 1562 psi does not exceed the allowable $F_b = 1600$ psi, the beam is safe with respect to strength in bending.

Experiments show that when notches are cut away to obtain a gradual change in cross section, as indicated in Fig. 18-1*c*, the loss in shearing strength is reduced. For such a condition the unit shearing stress is determined by using the actual depth of material above the notch and making no other reduction; for example, if the beam in the foregoing illustrative example has the notches shaped as shown in Fig. 18-1*c*, the unit horizontal shearing stress developed would be found from the regular shear formula $v = 3V/2bd$, using 8 in. for the value of d. Then, because the vertical end shear on the beam is 2308 lb (half the concentrated load plus half the beam weight),

$$v = \frac{3V}{2bd} = \frac{3 \times 2308}{2 \times 5.5 \times 8} = 79 \text{ psi}$$

Problem 18-1-A. A 4 × 12 beam, notched as shown in Fig. 18-1*b*, has a span of 10 ft. It carries a uniformly distributed load of 400 lb per lin ft including its own weight. If the depth of beam above the notch is 9 in., $F_v = 120$ psi, and $F_b = 1500$ psi, (1) is the beam safe with respect to horizontal shear? (2) Is the beam safe in bending?

Problem 18-1-B. A 4 × 12 beam has a span of 12 ft and is notched as shown in Fig. 18-1*b*. The depth of material above the notch is 8 in. Two loads of 1200 lb each are concentrated at the third points of the span. Stress data consist of $F_v = 95$ psi, $F_b = 1300$ psi. (1) Is the beam safe for horizontal shear? (2) Is the beam safe for bending?

18-2 Flitched Beams

Before the advent of glued-laminated wood beams, it was common practice to increase the strength of timber beams by the addition of steel plates. Two forms for such a reinforced beam section are shown in Fig. 18-2. This composite steel and wood member is known as a *flitched beam*. With little increase in the beam size, a considerable increase in strength is achieved by this

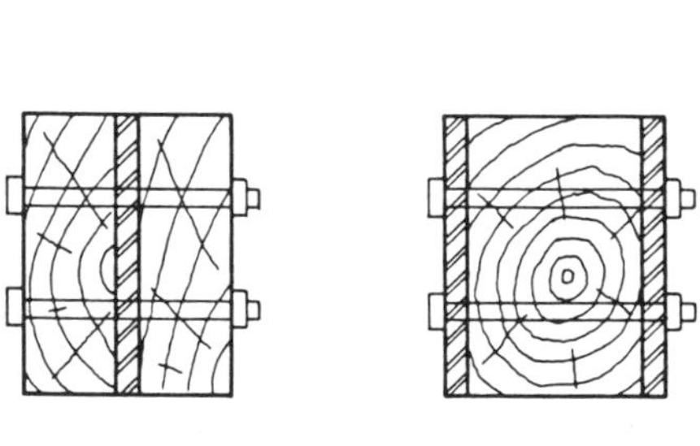

(a) (b)

FIGURE 18-2. Flitched beams.

technique. Of possible greater significance, however, is the increase in stiffness and the improvement of dimensional stability, especially with respect to long-term sag. Of course, the same can be accomplished through the use of a steel or glued-laminated beam, but where these are prohibitively expensive or not readily available the flitched beam is still in use.

The components of a flitched beam are securely held together with through bolts so that the elements act as a single unit. The computations for determining the strength of such a beam illustrate the phenomenon of two different materials in a beam acting as a unit. The computations are based on the premise that the two materials deform equally. Let

Δ_1 and Δ_2 = the deformations per unit length of the outermost fibers of the two materials, respectively,

f_1 and f_2 = the unit bending stresses in the outermost fibers of the two materials, respectively,

E_1 and E_2 = the moduli of elasticity of the two materials, respectively.

Since by definition the modulus of elasticity of a material is equal to the unit stress divided by the unit deformation, then

$$E_1 = \frac{f_1}{\Delta_1} \quad \text{and} \quad E_2 = \frac{f_2}{\Delta_2}$$

and transposing

$$\Delta_1 = \frac{f_1}{E_1} \quad \text{and} \quad \Delta_2 = \frac{f_2}{E_2}$$

Since the two deformations must be equal,

$$\frac{f_1}{E_1} = \frac{f_2}{E_2} \quad \text{and} \quad f_2 = f_1 \times \frac{E_2}{E_1}$$

This simple equation for the relationship between the stresses in the two materials of a composite beam may be used as the basis for investigation or design of a flitched beam, as is demonstrated in the following example.

Example. A flitched beam is formed as shown in Fig. 18-2*a* consisting of two 2 × 12 planks of Douglas fir, No. 1 grade, and a 0.5 × 11.25-in. [13 × 285 mm] plate of A36 steel. Compute the allowable uniformly distributed load this beam will carry on a simple span of 14 ft [4.2 m].

Solution: (1) We first apply the formula just derived to determine which of the two materials limits the beam action. For this we obtain the following data:

For the steel: E = 29,000,000 psi [200 GPa], and the maximum allowable bending stress F_b is 22 ksi [150 MPa].

For the wood: E = 1,800,000 psi [12.4 GPA], and the maximum allowable bending stress for single-member use is 1500 psi [10.3 MPa] (Table 3-1).

For a trial we assume the stress in the steel plate to be the limiting value and use the formula to find the maximum useable stress in the wood. Thus

$$f_w = f_s \times \frac{E_w}{E_s} = 22{,}000 \times \frac{1{,}800{,}000}{29{,}000{,}000} = 1366 \text{ psi [9.3 MPa]}$$

As this produces a stress lower than that of the table limit for the wood, our assumption is correct. That is, if we permit a stress higher than 1366 psi in the wood, the steel stress will exceed its limit of 22 ksi.

(2) Using the stress limit just determined for the wood, we now find the capacity of the wood members. Calling the load capacity of the wood W_w, we find

$$M = \frac{W_w L}{8} = \frac{W_w \times 14 \times 12}{8} = 21\, W_w$$

Then using the S of 31.6 in.^{3} for the 2 × 12 (Table 4-1), we find

$$M = 21\,W_w = f_w \times S_{2w} = 1366 \times (2 \times 31.6)$$

$$W_w = 4111 \text{ lb } [18.35 \text{ kN}]$$

(3) For the plate we first must find the section modulus as follows:

$$S_s = \frac{bd^2}{6} = \frac{0.5 \times (11.25)^2}{6} = 10.55 \text{ in.}^3\ [176 \times 10^3 \text{ mm}^3]$$

Then

$$M = 21\,W_s = f_s \times S_s = 22{,}000 \times 10.55$$

$$W_s = 11{,}052 \text{ lb } [50.29 \text{ kN}]$$

And the total capacity of the combined section is

$$W = W_w + W_s = 4111 + 11{,}052 = 15{,}163 \text{ lb } [68.64 \text{ kN}]$$

Although the load-carrying capacity of the wood elements is actually reduced in the flitched beam, the resulting total capacity is substantially greater than that of the wood members alone. This significant increase in strength achieved with small increase in size is a principal reason for popularity of the flitched beam. In addition, there is a significant reduction in deflection in most applications, and—most noteworthy—a reduction in sag over time.

Problem 18-2-A*. A flitched beam consists of a single 10 × 14 of Douglas fir, Select Structural grade, and two A36 steel plates, each 0.5 × 13.5 in. [13 × 343 mm] (See Fig. 18-2*b*.) Compute the magnitude of the concentrated load this flitched beam will support at the center of a 16 ft [4.8 m] simple span. Neglect the weight of the beam. Use a value of 22 ksi for the limiting bending stress in the steel.

18-3 Built-up Plywood and Lumber Elements

Various types of structural components can be produced with combinations of plywood and pieces of standard structural lum-

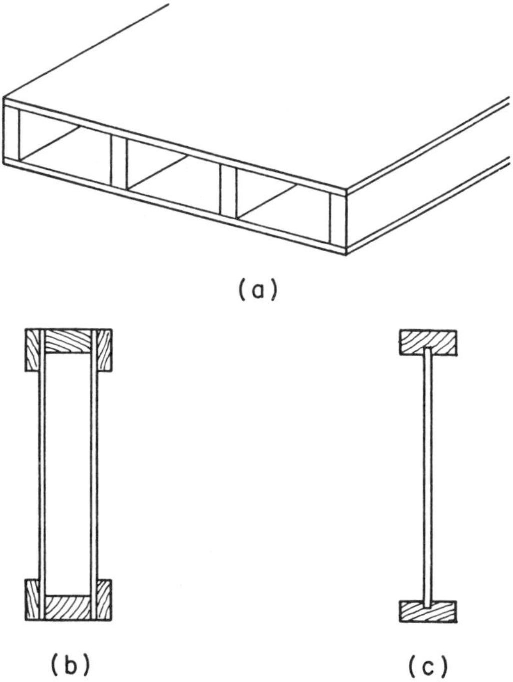

FIGURE 18-3. Composite built-up plywood and timber structural elements.

ber. Figure 18-3 shows some commonly used elements that can serve as structural components for buildings.

Stressed-skin Panels (Fig. 18-3*a*). This type of unit consists of two sheets of plywood attached to a core frame of solid-sawn elements. This is generally described as a plywood *sandwich panel*. However, when it is used in a manner that involves the development of spanning functions (as for a roof deck) and a true box-beam type of action is assumed, it is called a *stressed-skin panel*. These are used mostly for structures that have some modular, prefabricated system that may also involve the need for demountability for reuse. For ordinary situations, however, these panels are seldom competitive with typical construction utilizing plywood panels, joists, and the usual ceiling surfacing.

For the stressed-skin panel the plywood is usually glued to the wood frame, although if appearance is not a major concern, nails or screws may also be used. If gluing alone is used, it must be done in a factory with real quality control of the process.

The basic technique of construction for the simple unit shown in Fig. 18-3*a* may be extended to more complex forms such as curved panels or nonrectangular shapes. For special situations it may be possible to obtain plywood panels in sizes larger than the usual 4 by 8 ft so that sandwich panels can be fabricated with no face joints.

Some manufacturers produce these panels as standard products. Facings may vary for various applications and the unit voids may be filled with insulation for use as wall or roof construction.

Built-up Beams. In the spanning stressed-skin panel, tension and compression developed by bending are principally resisted by the top and bottom plywood faces, whereas shear is resisted by the lumber frame elements. These roles are reversed in the built-up beams shown in Figs. 18-3*b* and *c*. In these members the top and bottom lumber pieces function like the chords of a truss or the flanges of an I-beam, whereas the plywood web resists shear. These members are highly variable in response to different loads and span conditions.

This type of unit is also capable of variations in profile such as those illustrated for glued-laminated beams in Fig. 16-2. Webs can be penetrated with openings for the passage of wiring or piping in regions of low shear stress (usually not near the ends of the span). A great deal of customizing is possible; however, the widest usage is probably in the simple form shown in Fig. 18-3*c*, where a single web of plywood is glued into grooves cut in the single-piece wood flanges. This type of unit is used extensively for commercial buildings with roofs or floors of medium span range. Webs are usually of plywood, although beams with webs of particleboard are also produced.

18-4 Pole Structures

The straight slender trunks of various species of coniferous trees have been used for many structural purposes from ancient times to the present. Log cabins, pole stockades, fences, waterfront piers, and various utilitarian buildings have used poles in ways largely unchanged for many centuries. Modern developments consist mainly of more sophisticated connection devices and pressure treatment with chemicals for enhanced resistance to weather, insects, and vermin.

Poles may be used as beams or rafters, but are most often used as columns or for foundations. Timber piles are poles driven into the ground like large spikes by a pile driver. Another way to use poles for foundations, however, is by simply digging a hole, inserting the pole partway into the hole, and then filling the hole around the pole with soil or concrete—as is done for fence posts, sign supports, and utility poles. For a building the buried pole may be cut off a short distance above the ground and used to support a wood frame structure of ordinary construction, or the buried poles may extend upward to become columns for a frame structure.

Poles are usually produced by simply peeling the log of the bark and the soft inner layers on the outside of the tree trunk. This typically leaves a member that may be essentially straight, but has a tapered form and various surface irregularities such as knots, splits, and pitch pockets. Precision of form and perfect straightness are not realistically achieveable. Poles can be shaved to a truer round cross section and short untapered lengths are possible, but this adds to the cost and the appearance is less natural. One way to reduce the taper is to cut poles from a longer log instead of using a single tree trunk of the necessary length.

Poles tend to be used somewhat regionally where the climate, soil conditions, and availability of good cheap poles favor their selection. They can be used for simple utilitarian structures or for imaginative, sculptural building structures that are developed with a rough-textured and rustic timber style.

Design of timber piles, buried pole foundations, and round, tapered columns can be done with criteria from industry standards and many building codes. These guides may be used to develop a basis for structural computations, but much of the design of these structures is based on experience and the simple recognition of the fact that the structures have endured successfully for so long that it must be all right to build them that way.

18-5 Wood Fiber Products

Wood fiber products range from paper and cardboard to very dense and strong pressed hardboards. Actually the major volume of commercially forested wood goes into these products, mostly for newsprint, tissues, and containers. For structures, some current uses of wood fiber products are the following:

1. *Paper.* Paper is used extensively in building construction, although not much for primary structure. Drywall, or plasterboard, is actually a sandwich with a core of gypsum plaster and faces of paper; it is the number one interior wall-surfacing material. Stucco (exterior cement plaster) is often applied directly to a wood frame using a backup of only wire mesh bonded to paper. Paper may be coated, impregnated, or reinforced with various materials to improve its performance.
2. *Cardboard.* Cardboard is in essence merely very thick, stiff paper. A special form of cardboard is the familiar *corrugated cardboard* consisting of a sandwich with two flat paper faces and a corrugated interior core. Cardboard is not much used itself for building construction, but serves various purposes, such as the forming of some concrete systems. The corrugated sandwich is a model for a variety of products using a range of facings and core materials.
3. *Pressed Wood Fiber Panels.* Rigid, sheet-form panels of pressed wood fiber have been used for many years. These have seen wider use in nonstructural applications, but have recently steadily encroached in areas of use traditionally filled by plywood. Structural uses now include wall sheathing, roof decks, and the webs of built-up beams. (See Sec. 18-3 and Fig. 18-2*c*.)
4. *Composite Panel Products.* Various commercial building products are made with wood fibers (or other plant fibers) in combination with cement, asbestos fibers, asphalt, and so on. These may be used for essentially nonstructural purposes, such as roof or wall insulation, or may serve as

structural components in a frame structure—as wall surfacing or roof decks.

As in other situations with manufactured products, the primary source for design information is either industrywide organizations or individual manufacturers. As usage becomes widespread, building codes usually eventually have some criteria—as most now do for structural particleboard or hardboard.

Partly because they permit the use of wood materials that are not usable for plywood or structural lumber and partly because there is already a large wood fiber industry, it is likely that wood fiber products will see increasing use for structural applications. Trees, after all, are a renewable source as opposed to the raw material for most structural products.

19

Wood-Framed Diaphragms

Wood-framed stud walls, wood frame floor structures, and wood-framed, flat-surfaced roof structures are most often the principal elements used to brace buildings with wood structures against the lateral load effects of wind and earthquakes. When the wood skeleton frame is surfaced with a relatively stiff material, the resulting rigid planar elements are capable of spanning and of resisting shear distortion in their own planes. This in-plane action is called *diaphragm action* and the planar elements used for this structural function are called *diaphragms*. This chapter discusses the general nature and functioning of wood-framed diaphragms, principally those surfaced with structural grades of plywood. The materials presented here are condensed from a more extensive development of the topics in *Design for Lateral Forces* (Ref. 8).

19-1 Application of Wind and Seismic Forces

To understand how a building resists the lateral load effects of wind and seismic force it is necessary to consider the manner of application of the forces and then to visualize how these forces are transferred through the lateral resistive structural system and into the ground.

Wind Forces. The application of wind forces to a closed building is in the form of pressures applied normal to the exterior surfaces of the building. In one design method the total effect on the building is determined by considering the vertical profile, or silhouette, of the building as a single vertical plane surface at right angles to the wind direction. A direct horizontal pressure is assumed to act on this plane.

Figure 19-1 shows a simple rectangular building under the effect of wind normal to one of its flat sides. The lateral resistive structure that responds to this loading consists of the following:

Wall surface elements on the windward side are assumed to take the total wind pressure and are typically designed to span vertically between the roof and floor structures.

Roof and floor decks, considered as rigid planes (called diaphragms), receive the edge loading from the windward wall and distribute the load to the vertical bracing elements.

Vertical frames or shear walls, acting as vertical cantilevers, receive the loads from the horizontal diaphragms and transfer them to the building foundations.

The foundations must anchor the vertical bracing elements and transfer the loads to the ground.

The propagation of the loads through the structure is shown in the left part of Fig. 19-1 and the functions of the major elements of the lateral resistive system are shown in the right part of the figure. The exterior wall functions as a simple spanning element loaded by a uniformly distributed pressure normal to its surface and delivering a reaction force to its supports. In most cases, even though the wall may be continuous through several stories, it is considered as a simple span at each story level, thus delivering half of its load to each support. Referring to Fig. 19-1, this means that the upper wall delivers half of its load to the roof edge and half to the edge of the second floor. The lower wall delivers half of its load to the second floor and half to the first floor.

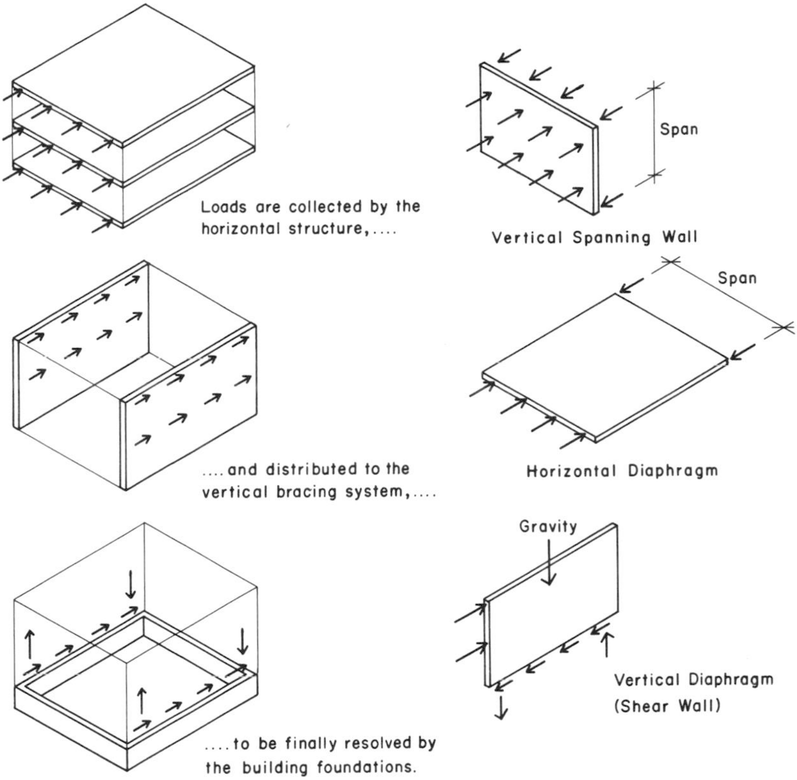
Loads are collected by the
horizontal structure,....
Span
Vertical Spanning Wall
....and distributed to the
vertical bracing system,....
Span
Horizontal Diaphragm
....to be finally resolved by
the building foundations.
Gravity
Vertical Diaphragm
(Shear Wall)

FIGURE 19-1. Propagation of wind force and basic functions of elements in a box building.

This may be a somewhat simplistic view of the function of the walls themselves, depending on their construction. If they are framed walls with windows or doors, there may be many internal load transfers within the wall. Usually, however, the external load delivery to the horizontal structure will be as described.

The roof and second floor diaphragms function as spanning elements loaded by the edge forces from the exterior walls and spanning between the end shear walls, thus producing a bending that develops tension on the leeward edge and compression on the windward edge. It also produces shear in the plane of the diaphragm that becomes a maximum at the end shear walls. In most cases the shear is assumed to be taken by the diaphragm, but the tension and compression forces due to bending are transferred to framing at the diaphragm edges. The means of achieving this transfer depends on the materials and details of the construction.

The end shear walls act as vertical cantilevers that also develop shear and bending. The total shear in the upper story is equal to the edge load from the roof. The total shear in the lower story is the combination of the edge loads from the roof and second floor. The total shear force in the wall is delivered at its base in the form of a sliding friction between the wall and its support. The bending caused by the lateral load produces an overturning effect at the base of the wall as well as the tension and compression forces at the edges of the wall. The overturning effect is resisted by the stabilizing effect of the dead load on the wall. If this stabilizing moment is not sufficient, a tension tie must be made between the wall and its support.

If the first floor is attached directly to the foundations, it may not actually function as a spanning diaphragm but rather will push its edge load directly to the leeward foundation wall. In any event, it may be seen in this example that only three-quarters of the total wind load on the building is delivered through the upper diaphragms to the end shear walls.

This simple example illustrates the basic nature of the propagation of wind forces through the building structure, but there are many other possible variations with more complex building forms or with other types of lateral resistive structural systems.

Seismic Forces. Seismic loads are actually generated by the dead weight of the building construction. In visualizing the application of seismic forces, we look at each part of the building and consider its weight as a horizontal force. The weight of the horizontal structure, although actually distributed throughout its plane, may usually be dealt with in a manner similar to the edge loading caused by wind. In the direction normal to their planes, vertical walls will be loaded and will function structurally in a manner similar to that for direct wind pressure. The load propagation for the box-shaped building in Fig. 19-1 will be quite similar for both wind and seismic forces.

If a wall is reasonably rigid in its own plane, it tends to act as a vertical cantilever for the seismic load in the direction parallel to its surface. Thus, in the example building, the seismic load for the roof diaphragm would usually be considered to be caused by the weight of the roof and ceiling construction plus only those walls whose planes are normal to the direction being considered.

For determination of the seismic load, it is necessary to consider all elements that are permanently attached to the structure. Ductwork, lighting and plumbing fixtures, supported equipment, signs, and so on, will add to the total dead weight for the seismic load. In buildings such as storage warehouses and parking garages it is also advisable to add some load for the building contents.

19-2 Horizontal Diaphragms

Most lateral resistive structural systems for buildings consist of some combination of vertical elements and horizontal elements. The horizontal elements are most often the roof and floor framing and decks. When the deck is of sufficient strength and stiffness to be developed as a rigid plane, it is called a *horizontal diaphragm*.

General Behavior A horizontal diaphragm typically functions by collecting the lateral forces at a particular level of the building and then distributing them to the vertical elements of the lateral resistive system. For wind forces the lateral loading of the horizontal diaphragm is usually through the attachment of the exterior wall to its edges. For seismic forces the loading is partly a result of the weight of the deck itself and partly a result of the weights of other parts of the building that are attached to it.

Relative Stiffness of the Horizontal Diaphragm. If the horizontal diaphragm is relatively flexible, it may deflect so much that its continuity is negligible and the distribution of load to the relatively stiff vertical elements is essentially on a peripheral basis. If the deck is quite rigid, on the other hand, the distribution to vertical elements will be essentially in proportion to their relative stiffness with respect to each other. The possibility of these two situations is illustrated for a simple box system in Fig. 19-2.

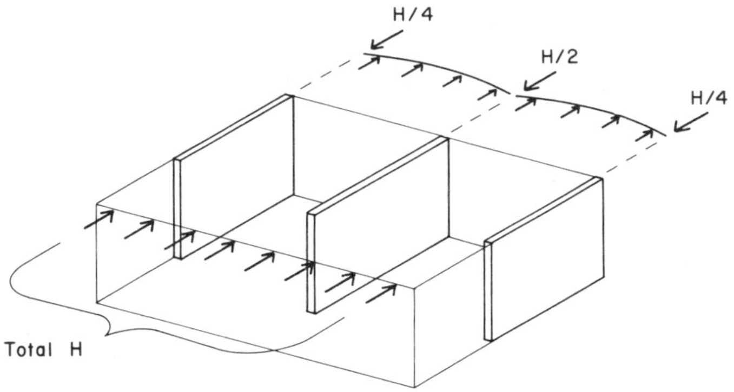

Peripheral distribution - flexible horizontal diaphragm

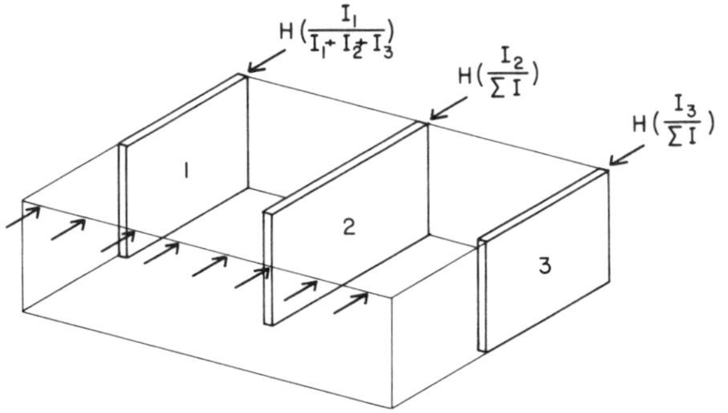

Proportionate stiffness distribution - rigid horizontal diaphragm

FIGURE 19-2. Distribution of forces by horizontal diaphragms.

Torsional Effects. If the centroid of the lateral forces in the horizontal diaphragm does not coincide with the centroid of the stiffness of the vertical elements, there will be a twisting action (called *rotation effect* or *torsional effect*) on the structure as well as the direct force effect. Figure 19-3 shows a structure in which this effect occurs because of a lack of symmetry of the structure. This effect is usually of significance only if the horizontal diaphragm is relatively stiff. This stiffness is a matter of the materials of the construction as well as the depth-to-span ratio of the hori-

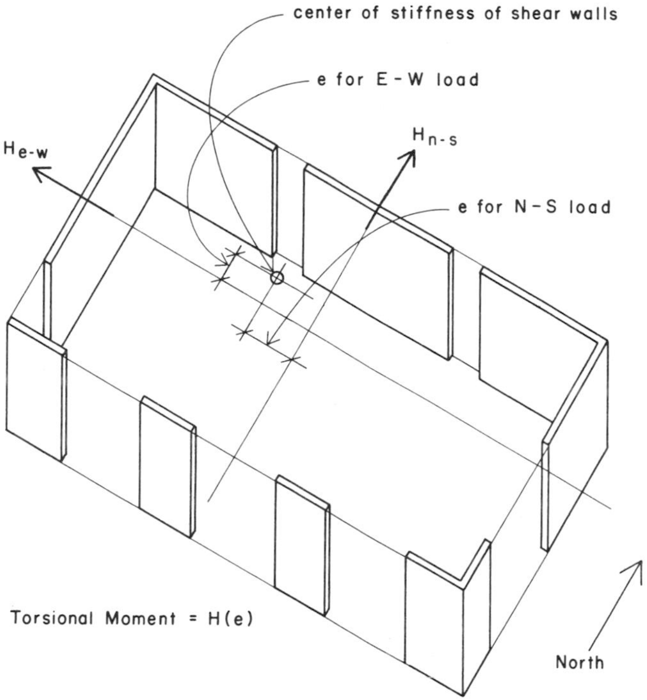

FIGURE 19-3. Torsional moment due to lateral forces in an unsymmetrical structure.

zontal diaphragm. In general, wood and metal decks are quite flexible, whereas poured concrete decks are very stiff.

Relative Stiffness of the Vertical Elements. When vertical elements share load from a rigid horizontal diaphragm, as shown in the lower figure in Fig. 19-2, their relative stiffness must usually be determined in order to establish the manner of the sharing. The determination is comparatively simple when the elements are similar in type and materials such as all plywood shear walls. When the vertical elements are different, such as a mix of plywood and masonry shear walls or of some shear walls and some braced frames, their actual deflections must be determined in order to establish the distribution, and this may require laborious computations.

Use of Control Joints. The general approach in design for lateral loads is to tie the whole structure together to assure its overall continuity of movement. Sometimes, however, because of the irregular form or large size of a building, it may be desirable to control its behavior under lateral loads by the use of structural separation joints. In some cases these joints function to create total separation, thus allowing for completely independent motion of the separate parts of the building. In other cases the joints may control movements in a single direction while achieving connection for load transfer in other directions.

Design and Usage Considerations. In performing their basic tasks, horizontal diaphragms have a number of potential stress problems. A major consideration is that of the shear stress in the plane of the diaphragm caused by the spanning action of the diaphragm as shown in Fig. 19-4. This spanning action results in shear stress in the material as well as a force that must be transferred across joints in the deck when the deck is composed of separate elements such as sheets of plywood or units of formed sheet metal. Figure 19-5 shows a typical plywood framing detail at the joint between two sheets. The stress in the deck at this location must be passed from one sheet through the edge nails to the framing member and then back out through the other nails to the adjacent sheet.

As is the usual case with shear stress, both diagonal tension and diagonal compression are induced simultaneously with the shear stress. The diagonal tension becomes critical in materials such as concrete. The diagonal compression is a potential source of buckling in decks composed of thin sheets of plywood or metal. In plywood decks the thickness of the plywood relative to the spacing of framing members must be considered, and it is also why the plywood must be nailed to intermediate framing members (not at edges of the sheets) as well as at edges.

Diaphragms with continuous deck surfaces are usually designed in a manner similar to that for webbed steel beams. The web (deck) is designed for the shear, and the flanges (edge-fram-

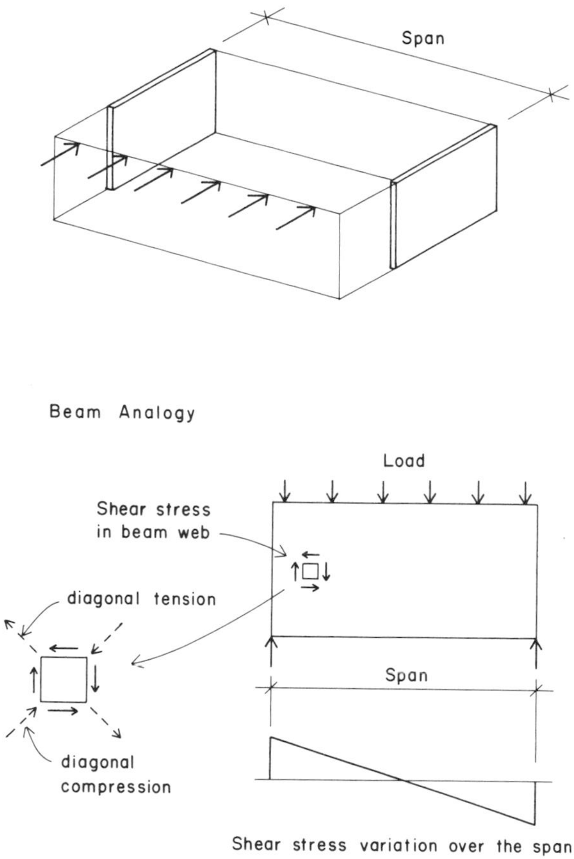

FIGURE 19-4. Beam functions of a horizontal diaphragm.

ing elements) are designed to take the moment, as shown in Figure 19-6. The edge members are called *chords,* and they must be designed for the tension and compression forces at the edges. With diaphragm edges of some length, the latter function usually requires that the edge members be spliced for some continuity of the forces. In many cases there are ordinary elements of the framing system, such as spandrel beams or top plates of stud walls, that have the potential to function as chords for the diaphragm.

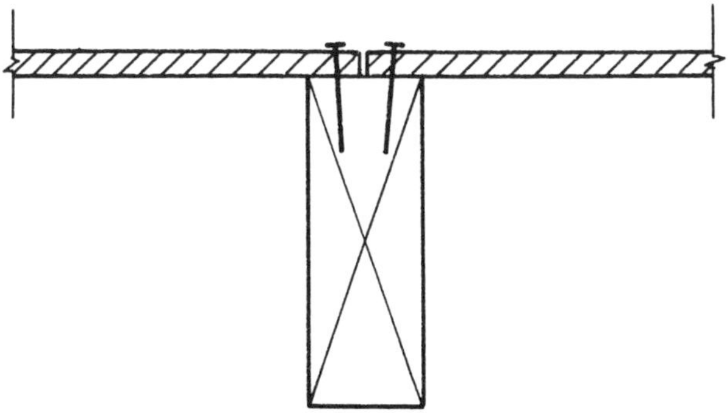

FIGURE 19-5. Typical nailed joint in a plywood diaphragm.

In some cases the collection of forces into the diaphragm or the distribution of loads to vertical elements may induce a stress beyond the capacity of the deck alone. Figure 19-7 shows a building in which a continuous roof diaphragm is connected to a series of shear walls. Load collection and force transfers require that some force be dragged along the dotted lines shown in the figure. For the outside walls it is possible that the edge framing used for chords can do double service for this purpose. For the interior shear wall, and possibly for the edges if the roof is cantilevered past the walls, some other framing elements may be necessary to reinforce the deck.

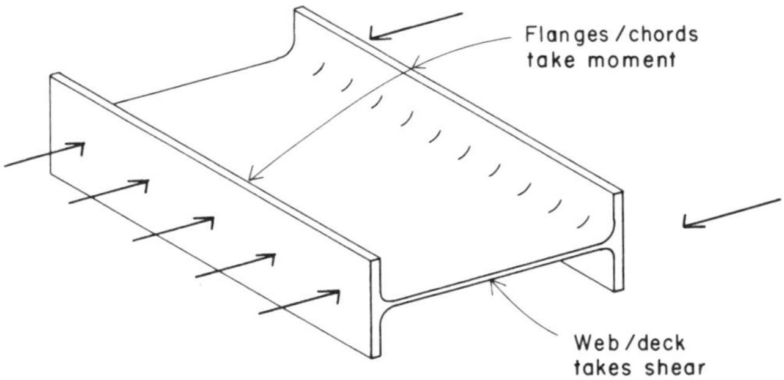

FIGURE 19-6. Flanged and webbed beam analogy for a horizontal diaphragm.

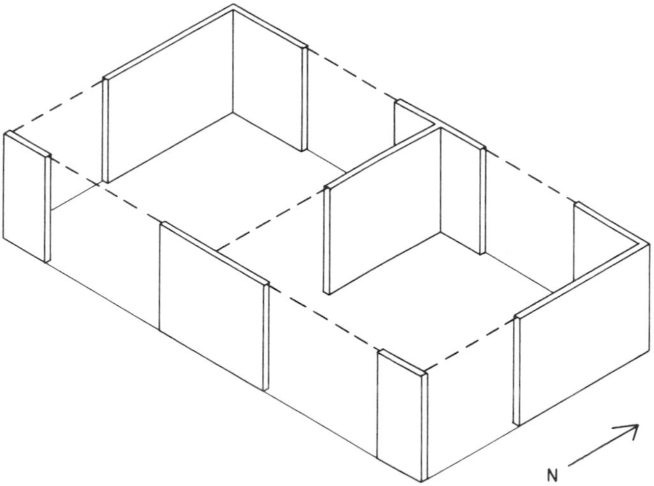

FIGURE 19-7. Collector elements in a shear wall braced structure.

Typical Construction. The most common horizontal diaphragm is the plywood deck for the simple reason that wood frame construction is so popular and plywood is mostly used for roof and floor decks. For roofs the deck may be as thin as $\frac{3}{8}$ in., but for flat roofs with waterproof membranes decks are usually $\frac{1}{2}$ in. or more. Attachment is typically by nailing, although glued floor decks are used for their added stiffness and to avoid nail popping and squeaking. Mechanical devices for nailing may eventually become so common that shear capacities will be based on some other fastener; at present the common wire nail is still the basis for load rating.

Attachment of plywood to chords and collectors and load transfers to vertical shear walls are also mostly achieved by nailing. Code-acceptable shear ratings are based on the plywood type and thickness, the nail size and spacing, and features such as size and spacing of framing and use of blocking. Load capacities for plywood decks are given in *Uniform Building Code,* Table 25-J, which is reprinted here as Table 19-1.

In general, plywood decks are quite flexible and should be investigated for deflection when spans are large or span–depth ratios are high.

TABLE 19-1. Allowable Shear in Pounds per Foot for Horizontal Plywood Diaphragms with Framing of Douglas Fir–Larch or Southern Pine[1]

PLYWOOD GRADE	Common Nail Size	Minimum Nominal Penetration in Framing (in Inches)	Minimum Nominal Plywood Thickness (in Inches)	Minimum Nominal Width of Framing Member (in Inches)	BLOCKED DIAPHRAGMS: Nail spacing at diaphragm boundaries (all cases), at continuous panel edges parallel to load (Cases 3 and 4) and at all panel edges (Cases 5 and 6): 6 / Nail spacing at other plywood panel edges: 6	4 / 6	2½[2] / 4	2[2] / 3	UNBLOCKED DIAPHRAGM: Nails spaced 6" max. at supported end: Load perpendicular to unblocked edges and continuous panel joints (Case 1)	Other configurations (Cases 2, 3 & 4)
STRUCTURAL I	6d	1¼	5/16	2	185	250	375	420	165	125
				3	210	280	420	475	185	140
	8d	1½	3/8	2	270	360	530	600	240	180
				3	300	400	600	675	265	200
	10d	1⅝	15/32	2	320	425	640	730[2]	285	215
				3	360	480	720	820	320	240
C-D, C-C, STRUCTURAL II and other grades covered in U.B.C. Standard No. 25-9	6d	1¼	5/16	2	170	225	335	380	150	110
				3	190	250	380	430	170	125
			3/8	2	185	250	375	420	165	125
				3	210	280	420	475	185	140
	8d	1½	3/8	2	240	320	480	545	215	160
				3	270	360	540	610	240	180
			15/32	2	270	360	530	600	240	180
				3	300	400	600	675	265	200
	10d	1⅝	15/32	2	290	385	575	655[2]	255	190
				3	325	430	650	735	290	215
			19/32	2	320	425	640	730[2]	285	215
				3	360	480	720	820	320	240

[1]These values are for short-time loads due to wind or earthquake and must be reduced 25 percent for normal loading. Space nails 10 inches on center for floors and 12 inches on center for roofs along intermediate framing members.

Allowable shear values for nails in framing members of other species set forth in Table No. 25-17-J of U.B.C. Standards shall be calculated for all grades by multiplying the values for nails in STRUCTURAL I by the following factors: Group III, 0.82 and Group IV, 0.65.

[2]Framing shall be 3-inch nominal or wider and nails shall be staggered where nails are spaced 2 inches or 2½ inches on center, and where 10d nails having penetration into framing of more than 1⅝ inches are spaced 3 inches on center.

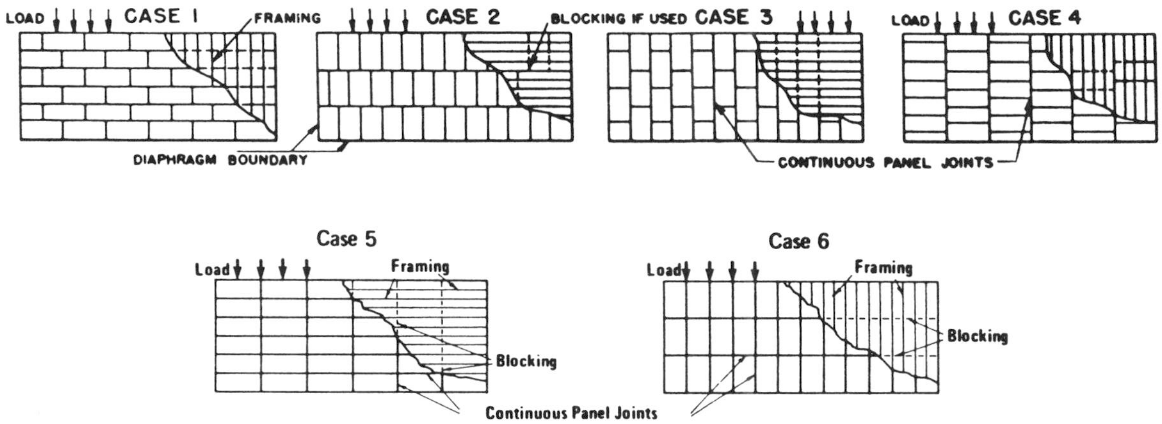

NOTE: Framing may be located in either direction for blocked diaphragms.

Source: This is Table 25-J-1 from the *Uniform Building Code*, 1985 edition, reprinted with permission of the publishers, International Conference of Building Officials.

Decks of boards or timber, usually with tongue-and-groove joints, were once popular but are given low rating for shear capacity at present. Where the exposed plank-type deck is desired, it is not uncommon to use a thin plywood deck on top of it for lateral force development.

Many other types of roof deck construction may function adequately for diaphragm action, especially when the required unit shear resistance is low. Acceptability by local building code administration agencies should be determined if any construction other than those described is to be used.

Shear Capacity of Plywood Decks. Table 19-1 yields the capacity of plywood decks in units of pounds per foot of deck width. The table incorporates a number of variables as follows:

1. *Arrangement of the Plywood Panels.* The table footnotes indicate six cases of panel arrangement as related to the direction of the load. Data in the table are limited in some cases based on these panel layouts.
2. *Provision of Blocking.* This refers to the use of extra framing provided between rafters or joists to support the deck edges and allow for nailing. Omission of blocking is permitted only with certain panel arrangements and in general results in lowered shear capacities.
3. *Type of Plywood.* The two general classes of plywood indicated are Structural I and Structural II, with various other grades grouped with Structural II.
4. *Size of Nail.* This is generally related to the thickness of the plywood; nails are ordinary common wire nails, although a somewhat shorter special plywood nail is often used.
5. *Thickness of Plywood.* Table values are given for thicknesses ranging from $\frac{5}{16}$ in. to $\frac{19}{32}$ in.
6. *Size of Framing.* Loads are given for nominal 2 in. and 3 in. framing.
7. *Special Considerations.* Table footnotes provide for additional modifications in some cases.

Other building code provisions require a minimum nail spacing of 6 in. on center at all panel edges and 12 in. on center at points of support within the panel (called *field nailing*). The following examples illustrate the use of the data in Table 19.1.

Example 1. Determine the maximum shear capacity for a horizontal diaphragm consisting of plywood panels nailed to 2 in. nominal framing of Douglas fir–larch. Data for the construction are as follows:

Panels are Structural II, $\frac{15}{32}$ in. thick, arranged in the Case 2 pattern and attached to blocked framing.

Nails are 8d spaced 4 in. at diaphragm boundaries and 6 in. at all other panel edges.

Solution: For this combination of conditions the table yields a value of 360 lb/ft.

Example 2. A roof deck made with $\frac{15}{32}$ in. Structural I plywood is required to resist a diaphragm shear loading that results in a stress of 450 lb/ft in the deck. The blocked diaphragm has the plywood panels as shown for layout Case 1. Find the nail size and spacing required for (1) 2 in. nominal framing and (2) 3 in. nominal framing.

Solution: (1) For the 2 in. framing, the required spacing for the 10d nails is $2\frac{1}{2}$ in. at the diaphragm boundaries and 4 in. at other panel edges. However, reference to the footnote in Table 19-1 indicates that for the $2\frac{1}{2}$ in. spacing the framing must be 3 in. nominal. Therefore the 2 in. framing cannot be used to achieve this level of shear capacity.

(2) If the 3 in. framing is used, the table indicates a shear capacity of 480 lb/ft with nails at 4 in. at diaphragm boundaries and at 6 in. at other panel edges.

Example 3. A one-story wood-framed building has the form shown in Fig. 19-8. The roof consists of wood framing and plywood deck with $\frac{15}{32}$ in. minimum thickness required for roofing installation. For lateral load resistance the roof must develop a unit shear of 381 lb/ft at the ends of the building. Shear stress in general is distributed as for a beam with uniformly distributed

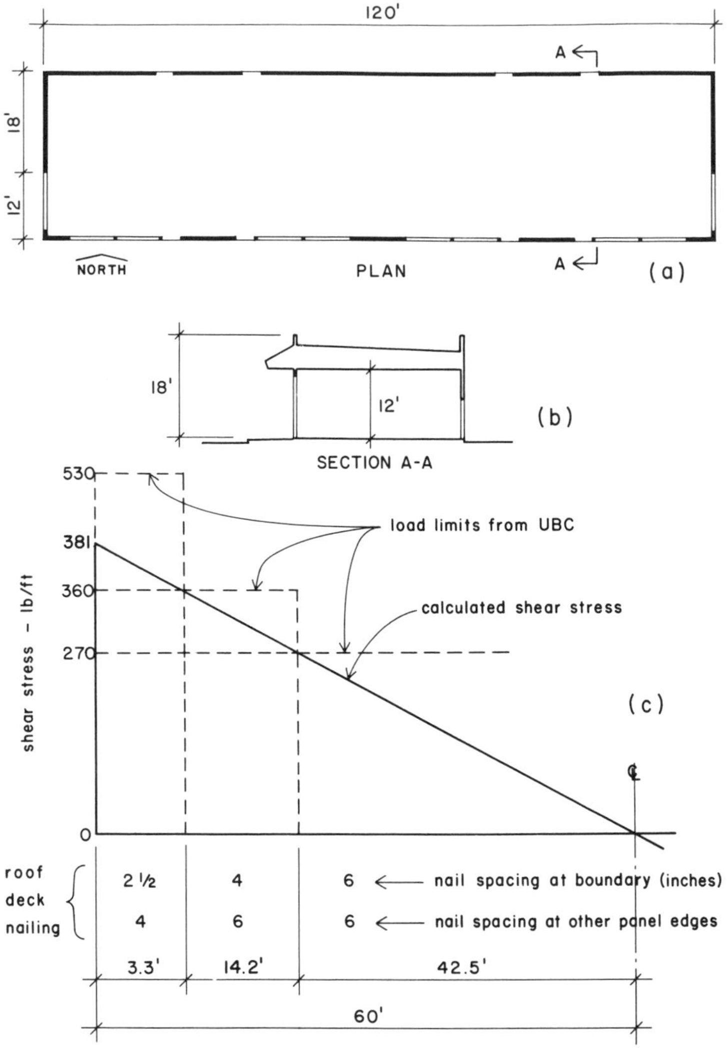

FIGURE 19-8. Example 3.

loading, varying from the maximum value at the ends to zero at midspan. Determine options for the roof construction based only on considerations of lateral effects.

Solution: We will assume a Case 1 panel layout, as this is optimal for a high shear condition. For the maximum shear stress, some options from Table 19-1 are the following:

$\frac{15}{32}$ in. Structural II with $2\times$ framing, blocking, and 8d nails at $2\frac{1}{2}$ in. at boundaries and at 4 in. at other edges.

$\frac{19}{32}$ in. Structural II with $3\times$ framing, blocking, and 10d nails at 4 in. at boundaries and at 6 in. at other edges.

$\frac{15}{32}$ in. Structural I with $3\times$ framing, blocking, and 10d nails at 4 in. at boundaries and at 6 in. at other edges.

Because the high stresses occur only near the ends of the building, it is reasonable to consider the possibility of zoning the deck nailing in this case. By using the $\frac{15}{32}$ in. Structural II plywood, it is possible to use two fewer nail spacings, as given in the *Uniform Building Code* table. Thus the range of nail spacings and corresponding load ratings are the following:

8d at $2\frac{1}{2}$ in. at boundaries, 4 in. at other edges: load = 530 lb/ft.
8d at 4 in. at boundaries, 6 in. at other edges: load = 360 lb/ft.
8d at 6 in. at all edges: load = 270 lb/ft.

In Fig. 19-8*c* these allowable loads are plotted on the graph of stress variation in the deck to permit the determination of the areas in which the various nailing spacings are usable. The maximum nailing is seen to be required on only a very small area at each end of the roof. The actual dimensions of the specified nailing zones may be adjusted slightly to correspond to modules of the roof framing and plywood sheet layouts as long as the limits of the calculated zone boundaries are not exceeded.

19-3 Vertical Diaphragms

Vertical diaphragms are usually the walls of buldings. As such, in addition to their shear wall function, they must fulfill various architectural functions and may also be required to serve as bearing walls for the gravity loads. The location of walls, the materials used, and some of the details of their construction must be developed with all these functions in mind.

The most common shear wall constructions are those of poured concrete, masonry, and wood frames of studs with surfacing elements. Wood frames may be made rigid in the wall plane by the use of diagonal bracing or by the use of surfacing materials that have sufficient strength and stiffness. Choice of the type of construction may be limited by the magnitude of shear caused by the lateral loads, but will also be influenced by fire code requirements and the satisfaction of the various other wall functions, as described previously.

General Behavior. Some of the structural functions usually required of vertical diaphragms are the following (see Fig. 19-9):

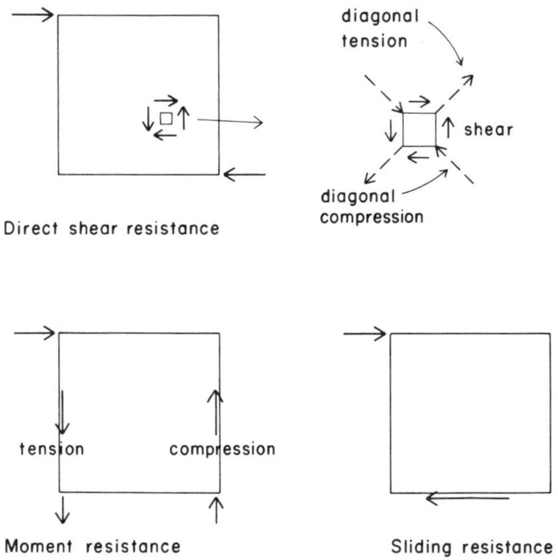

FIGURE 19-9. Functions of vertical diaphragms (shear walls).

1. *Direct Shear Resistance.* This usually consists of the transfer of a lateral force in the plane of the wall from some upper level of the wall to a lower level or to the bottom of the wall. This results in the typical situation of shear stress and the accompanying diagonal tension and compression stresses, as discussed for horizontal diaphragms.
2. *Cantilever Moment Resistance.* Shear walls generally work like vertical cantilevers, developing compression on one edge and tension on the opposite edge, and transferring an overturning moment to the base of the wall.
3. *Horizontal Sliding Resistance.* The direct transfer of the lateral load at the base of the wall produces the tendency for the wall to slip horizontally off its supports.

The shear stress function is usually considered independently of other structural functions of the wall. The maximum shear stress that derives from lateral loads is compared to some rated capacity of the wall construction, with the usual increase of one-third in allowable stresses because the lateral load is most often a result of wind or earthquake forces. For structurally surfaced wood frames the construction as a whole is generally rated for its total resistance in pounds per foot of the wall length in plan. For a plywood-surfaced wall this capacity depends on the type and thickness of the plywood; the size, wood species, and spacing of the studs; the size and spacing of the plywood nails; and the inclusion or omission of blocking at horizontal plywood joints.

For wood stud walls the *Uniform Building Code* provides tables of rated load capacities for several types of surfacing including plywood, diagonal wood boards, plaster, gypsum drywall, and particleboard.

Although the possibility exists for the buckling of walls as a result of the diagonal compression effect, this is usually not critical because other limitations exist to constrain wall slenderness. Slenderness of wood studs is limited by gravity design and by the code limits as a function of the stud size. Because stud walls are usually surfaced on both sides, the resulting sandwich–panel effect is usually sufficient to provide a reasonable stiffness.

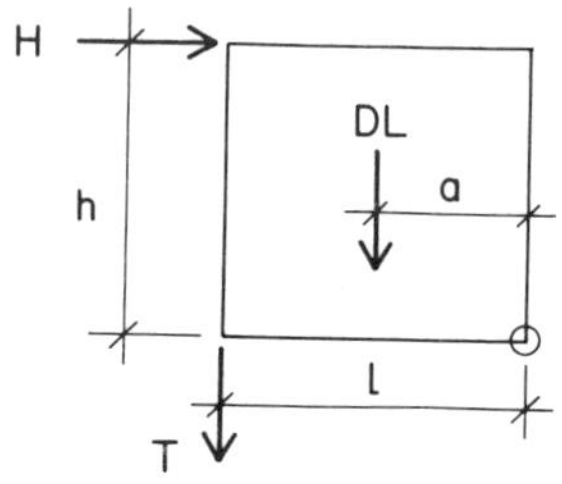

To determine T:

$$DL(a) + T(l) = 1.5\,[H(h)]$$

FIGURE 19-10. Overturn analysis for a vertical diaphragm.

As in the case of horizontal diaphragms, the moment effect on the wall is usually considered to be resisted by the two vertical edges of the wall acting as flanges or chords. In wood-framed walls the end-framing members are considered to fulfill this function. These edge members must be investigated for possible critical combinations of loading because of gravity and the lateral effects.

The overturn effect of the lateral loads should be resisted with a safety factor of 1.5 that is required by the *Uniform Building Code*. The form of the analysis for the overturn effect is as shown in Fig. 19-10. If the tiedown force is actually required, it is developed by the anchorage of the edge-framing elements of the wall.

Resistance to horizontal sliding at the base of a shear wall is usually at least partly resisted by friction caused by the dead loads. For wood-framed walls the friction is usually ignored and the sill bolts are designed for the entire load.

Design and Usage Considerations. An important judgment that must often be made in designing for lateral loads is that of the manner of distribution of lateral force between a number of shear walls that share the load from a single horizontal diaphragm. In some cases the existence of symmetry or of a flexible horizontal diaphragm may simplify this consideration. In many cases, however, the relative stiffnesses of the walls must be determined for this calculation.

If considered in terms of static force and elastic stress–strain conditions, the relative stiffness of a wall is inversely proportionate to its deflection under a unit load. Figure 19-11 shows the manner of deflection of a shear wall for two assumed conditions. In (*a*) the wall is considered to be fixed at its top and bottom, flexing in a double curve with an inflection point at midheight. This is the case usually assumed for a continuous wall of concrete or masonry in which a series of individual wall portions (called *piers*) are connected by a continuous upper wall or other structure of considerable stiffness. In (*b*) the wall is considered to be

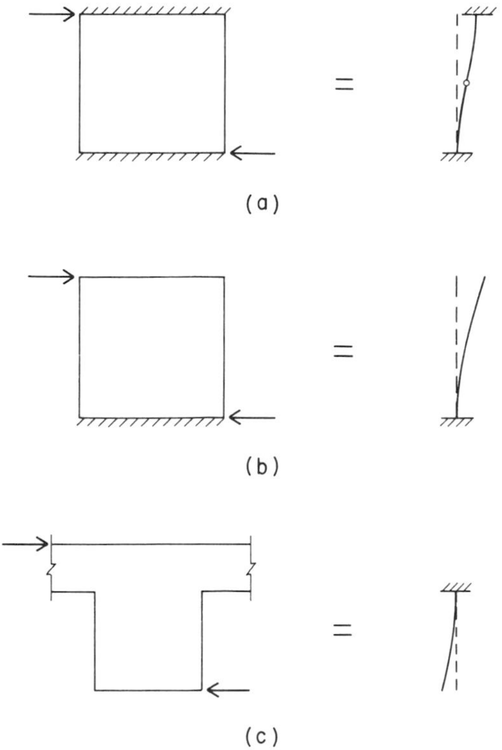

FIGURE 19-11. Alternative support and deformation behavior for isolated shear walls (also called piers).

fixed at its bottom only, functioning as a vertical cantilever. This is the case for independent, freestanding walls or for walls in which the continuous upper structure is relatively flexible. A third possibility is shown in (*c*) in which relatively short piers are assumed to be fixed at their tops only, which produces the same deflection condition as in (*b*).

In some instances the deflection of the wall may result largely from shear distortion rather than from flexural distortion, perhaps because of the wall materials and construction or the proportion of wall height to plan length. Furthermore, stiffness in resistance to dynamic loads is not quite the same as stiffness in resistance to static loads. The following recommendations are made for single-story shear walls:

1. For wood-framed walls with height-to-length ratios of 2 or less, assume the stiffness to be proportional to the plan length of the wall.
2. For wood-framed walls with height-to-length ratios over 2 and for concrete and masonry walls, assume the stiffness to be a function of the height-to-length ratio and the method of support (cantilevered or fixed top and bottom).
3. Avoid situations in which walls of significantly great differences in stiffness share loads along a single row. The short walls will tend to receive a small share of the loads, especially if the stiffness is assumed to be a function of the height-to-length ratio.
4. Avoid mixing of shear walls of different construction when they share loads on a deflection basis.

Item 4 in the preceding list can be illustrated by two situations as shown in Fig. 19-12. The first situation is that of a series of panels in a single row. If some of these panels are of concrete or masonry and others of wood frame construction, the stiffer concrete or masonry panels will tend to absorb the major portion of the load. The load sharing must be determined on the basis of actual calculated deflections. Better yet is a true dynamic analysis, because if the load is truly dynamic in character, the periods of the two types of walls are of more significance than their stiffness.

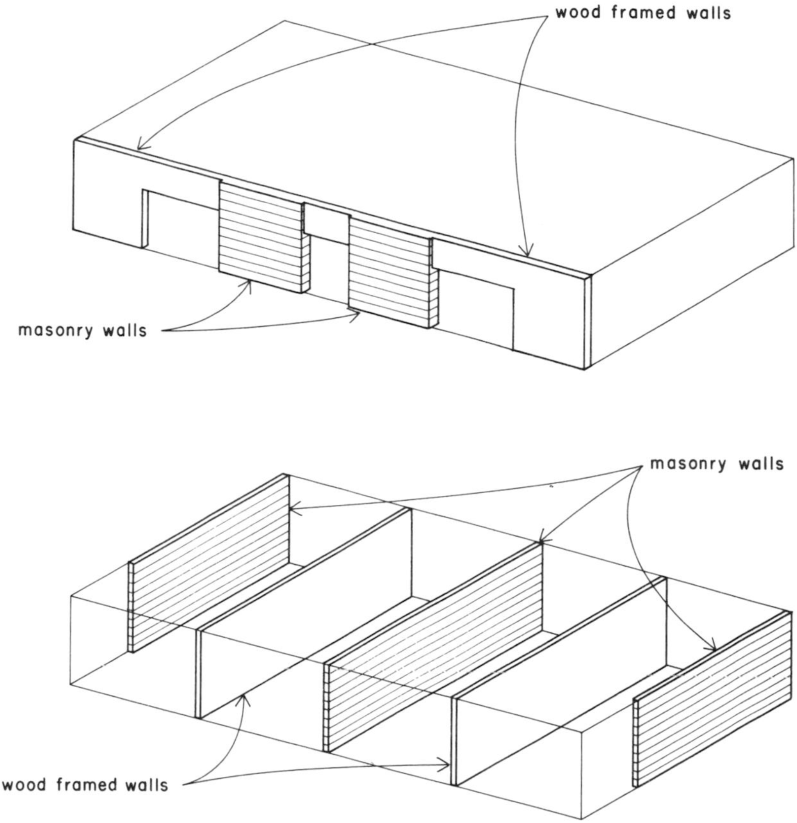

FIGURE 19-12. Interactive functioning of shear walls of mixed construction.

In the second situation shown in Fig. 19-12, the walls share load from a rigid horizontal diaphragm. This situation also requires a deflection calculation for determining the distribution of force to the panels.

In addition to the various considerations mentioned for the shear walls themselves, care must be taken to assure that they are properly anchored to the horizontal diaphragms.

A final consideration for shear walls is that they must be made an integral part of the whole building construction. In long build-

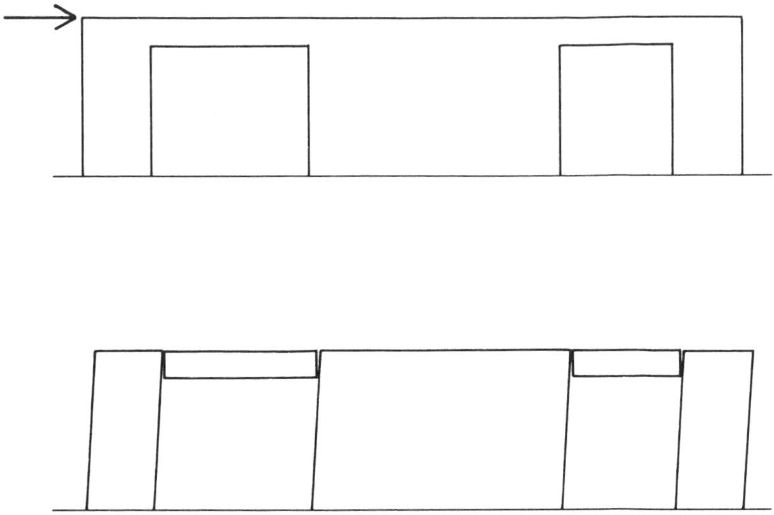

FIGURE 19-13. Effects of shear wall deformation.

ing walls with large door or window openings or other gaps in the wall, shear walls are often considered as entities (isolated, independent piers) for their design. However, the behavior of the entire wall under lateral load should be studied to be sure that elements not considered to be parts of the lateral resistive system do not suffer damage because of the wall distortions.

An example of this situation is shown in Fig. 19-13. The two relatively long solid portions are assumed to perform the bracing function for the entire wall and would be designed as isolated piers. However, when the wall deflects, the effect of the movement on the shorter piers, on the headers over openings, and on the door and window framing must be considered. The headers must not be cracked loose from the solid wall portions or pulled off their supports.

Typical Construction. The various types of common construction for shear walls mentioned in the preceding section are wood frames with various surfacing, reinforced masonry, and concrete. The only wood frame wall used extensively in the past was the plywood covered one. Experience and testing have established acceptable ratings for other surfacing, so that plywood is used somewhat less when shear loads are low.

For all types of walls there are various considerations (good carpentry, fire resistance, available products, etc.) that establish a certain minimum construction. In many situations this "minimum" is really adequate for low levels of shear loading, and the only additions are in the area of attachments and joint load transfers. Increasing wall strength beyond the minimum usually requires increasing the size or quality of units, adding or strengthening attachments, developing supporting elements to function as chords or collectors, and so on. It well behooves the designer to find out the standards for basic construction to know what the minimum consists of so that added strength can be developed when necessary—but using methods consistent with the ordinary types of construction.

19-4 Investigation and Design of Plywood Shear Walls

Plywood shear walls occur most often as exterior building walls with the shear-resisting plywood attached to the outside surface of the wall framing. Long building walls that are interrupted by openings for doors and windows are usually designed as a series of individual linked piers, with the piers consisting of the solid walls between openings. Distribution of the lateral force to an individual pier depends on many factors that include considerations of the overall building form and various details of the construction. Some typical situations are illustrated in Chapter 20.

A primary consideration is that for the shear stress in the plywood surface. Table 19-2 yields the capacity of plywood shear walls with ordinary wood stud framing and typical structural grades of Douglas fir plywood. The table incorporates a number of variables as follows:

1. *Plywood Grade*. Table 19-2 gives values for three grades, as classified by the *Uniform Building Code* designations. Other codes may use different terminology, although the basic data come from industrywide standards.

TABLE 19-2. Allowable Shear for Wind or Seismic Forces in Pounds per Foot for Plywood Shear Walls with Framing of Douglas Fir–Larch or Southern Pine[1,4]

PLYWOOD GRADE	MINIMUM NOMINAL PLYWOOD THICKNESS (Inches)	MINIMUM NAIL PENETRATION IN FRAMING (Inches)	NAIL SIZE (Common or Galvanized Box)	PLYWOOD APPLIED DIRECT TO FRAMING — Nail Spacing at Plywood Panel Edges: 6	4	3	2[2]	NAIL SIZE (Common or Galvanized Box)	PLYWOOD APPLIED OVER ½-INCH GYPSUM SHEATHING — Nail Spacing at Plywood Panel Edges: 6	4	3	2[2]
STRUCTURAL I	5/16	1¼	6d	200	300	390	510	8d	200	300	390	510
	3/8	1½	8d	230[3]	360[3]	460[3]	610[3]	10d	280	430	550[2]	730[2]
	15/32	1½	8d	280	430	550	730	10d	280	430	550[2]	730
	15/32	1⅝	10d	340	510	665[2]	870	—	—	—	—	—
C-D, C-C STRUCTURAL II and other grades covered in U.B.C. Standard No. 25-9.	5/16	1¼	6d	180	270	350	450	8d	180	270	350	450
	3/8	1¼	6d	200	300	390	510	8d	200	300	390	510
	3/8	1½	8d	220[3]	320[3]	410[3]	530[3]	10d	260	380	490[2]	640
	15/32	1½	8d	260	380	490	640	10d	260	380	490[2]	640
	15/32	1⅝	10d	310	460	600[2]	770	—	—	—	—	—
	19/32	1⅝	10d	340	510	665[2]	870	—	—	—	—	—
			NAIL SIZE (Galvanized Casing)					NAIL SIZE (Galvanized Casing)				
Plywood panel siding in grades covered in U.B.C. Standard No. 25-9.	5/16	1¼	6d	140	210	275	360	8d	140	210	275	360
	3/8	1½	8d	130[3]	200[3]	260[3]	340[3]	10d	160	240	310[2]	410

[1]All panel edges backed with 2-inch nominal or wider framing. Plywood installed either horizontally or vertically. Space nails at 6 inches on center along intermediate framing members for 3/8-inch plywood installed with face grain parallel to studs spaced 24 inches on center and 12 inches on center for other conditions and plywood thicknesses. These values are for short-time loads due to wind or earthquake and must be reduced 25 percent for normal loading.

Allowable shear values for nails in framing members of other species set forth in Table No. 25-17-J of U.B.C. Standards shall be calculated for all grades by multiplying the values for common and galvanized box nails in STRUCTURAL I and galvanized casing nails in other grades by the following factors: Group III, 0.82 and Group IV, 0.65.

[2]Framing shall be 3-inch nominal or wider and nails shall be staggered where nails are spaced 2 inches on center, and where 10d nails having penetration into framing of more than 1 5/8 inches are spaced 3 inches on center.

[3]The values for 3/8-inch-thick plywood applied direct to framing may be increased 20 percent, provided studs are spaced a maximum of 16 inches on center or plywood is applied with face grain across studs.

[4]Where plywood is applied on both faces of a wall and nail spacing is less than 6 inches on center on either side, panel joints shall be offset to fall on different framing members or framing shall be 3-inch nominal or thicker and nails on each side shall be staggered.

Source: This is Table 25-K-1 from the *Uniform Building Code,* 1985 edition, reprinted with permission of the publishers, International Conference of Building Officials.

2. *Thickness of Plywood.* A range of thicknesses is shown, although the most widely used are $\frac{3}{8}$ and $\frac{15}{32}$ in. panels.
3. *Nail Size.* This is associated with the plywood thickness; the smallest nails possible are preferred because of ease of installation and less likelihood of splitting of the studs. The table is based on ordinary wire (common) nails, although various power-driven fasteners are used when code approved for equivalent strength.
4. *Nail Spacing.* Maximum spacing at panel edges is specified elsewhere in the code at 6 in. Data are given for spacings down to 2 in., although this close spacing is seldom used because the large number of nails and splitting of the studs present installation difficulties. Studs wider than 2 in. nominal are required for close spacing or large nails. As with roof and floor decks, the interior portion of the panels is attached with nails at a maximum of 12 in. on center at all supports.
5. *Stud Spacing.* For thin panels the buckling of the plywood may be a problem, which results in some restrictions on stud spacing.
6. *Method of Application of the Plywood.* Plywood is usually attached directly to the studs. However, Table 19-2 also gives values for a special construction in which the plywood is applied over a $\frac{1}{2}$ in. thick gypsum sheathing.
7. *Special Considerations.* Table footnotes give some special restrictions and allowances for special situations.

In addition to consideration of shear, a wall must be investigated for the various other functions involved in its use, including the other shear wall functions described in the preceding sections. The following examples illustrate typical procedures for investigation and design of individual shear walls. Broader consideration of wall functions is illustrated in Chapter 20.

Example 1. A plywood shear wall having the form shown in Fig. 19-14 is required to resist the lateral force indicated. Assuming the lateral force to be due to wind, and the total dead weight on

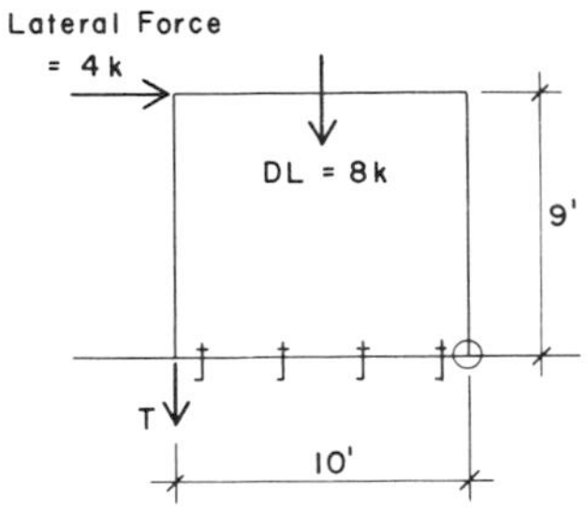

FIGURE 19-14.

the wall to be as shown, design the wall and its framing for the following data:

Plywood: Douglas fir, Structural II grade.
Wall framing: Douglas fir–larch, No. 2 or Stud grade.
Wall sill-bolted to a concrete base.

Solution: We first consider the maximum unit shear in the plywood as follows:

$$v = \frac{\text{lateral force}}{\text{wall length}} = \frac{4000}{10} = 400 \text{ lb/ft}$$

Assuming the plywood to be applied directly to the studs, we determine from Table 19-2 the following options for the Structural II plywood:

$\frac{3}{8}$ in. plywood, 6d nails at 2 in. (v = 510 lb/ft),
$\frac{3}{8}$ in. plywood, 8d nails at 3 in. (v = 410 lb/ft),
$\frac{15}{32}$ in. plywood, 10d nails at 4 in. (v = 460 lb/ft).

Note that table footnote 3 permits an increase of 20% in the value for the $\frac{3}{8}$ in. plywood with 8d nails under certain circumstances, although this is not critical for this example. Choice of the plywood and its nailing must be done with consideration of the general details of the construction and the other functions of the wall. For the given conditions of this example, any of the three options is adequate.

For the overturn analysis, as illustrated in Fig. 19-10, we determine the following:

$$\begin{aligned}
\text{Overturning moment} &= (\text{lateral force}) \times (\text{wall height}) \\
&\qquad \times (\text{safety factor}) \\
&= (4)(9)(1.5) \\
&= 54 \text{ k-ft} \\
\text{Restoring moment} &= (\text{dead weight}) \times (\tfrac{1}{2} \text{ wall length}) \\
&= (8)\left(\frac{10}{2}\right) \\
&= 40 \text{ k-ft}
\end{aligned}$$

Because the overturning effect (with safety factor) prevails, a tiedown force is required at the wall end, with its value equal to

$$\begin{aligned}
T &= \frac{(\text{overturning moment}) - (\text{restoring moment})}{\text{wall length}} \\
&= \frac{54 - 40}{10} = 1.4 \text{ kip}
\end{aligned}$$

The usual provision for this anchorage consists of a steel device bolted to the end framing of the wall and connected to a large anchor bolt embedded in the concrete base. Various patented devices, available as stock hardware items, may be used for this purpose. The force of 1.4 kip is substantial but within the feasible range of available devices and the capacity of ordinary wall framing.

The usual minimum sill bolting required by building codes is for one bolt not over 12 in. from each end of the wall and additional bolts not over 6 ft on center. This would result in a minimum of three bolts for this wall, usually with a minimum $\frac{1}{2}$ in.

diameter. Based on single shear in a sill of nominal 2 in. thickness, Table 13-2 yields a value of 470 lb for one bolt. With three bolts and an allowable stress increase of one-third for wind, the total capacity of the minimum bolting is thus

$$H = 3 \times 470 \times 1.333 = 1880 \text{ lb}$$

Because this is considerably short of the required 4000 lb, some increase is required over the minimum bolting. Possible choices include more bolts, larger bolts, a dense grade sill, or a thicker sill. Again, other factors of the general construction must be considered for a real situation. The least desirable choice is probably for a large number of bolts because setting in the concrete is a major labor cost.

20

Building Design Examples

This chapter presents illustrations of the design of the wood structures for two buildings. Design of individual elements of the structural systems is based on the materials presented in the earlier chapters. The principal purpose here is to show the broader context of design work by dealing with whole structures.

20-1 Introduction

Materials, methods, and details of building construction vary considerably on a regional basis. There are many factors that affect this situation, including the real effects of response to climate and the availability of construction materials. Even in a single region, differences occur between individual buildings, based on individual styles of architectural design and personal techniques of builders. Nevertheless, at any given time there are usually a few predominant, popular methods of construction that are employed for most buildings of a given type and size. The construction methods and details shown here are reasonable, but in no way are they intended to illustrate a singular, superior style of building.

20-2 Dead Loads

Dead load consists of the weight of the materials of which the building is constructed such as walls, partitions, columns, framing, floors, roofs, and ceilings. In the design of a beam, the dead load must include an allowance for the weight of the beam itself. Table 20-1, which lists the weights of many construction materials, may be used in the computation of dead loads. Dead loads are due to gravity and they result in downward vertical forces.

20-3 Roof Loads

In addition to the dead loads they support, roofs are designed for a uniformly distributed live load that includes snow accumulation and the general loadings that occur during construction and maintenance of the roof. Snow loads are based on local snowfalls and are specified by local building codes.

Table 20-2 gives the minimum roof live load requirements specified by the 1985 edition of the *Uniform Building Code*. Note the adjustments for roof slope and for the total area of roof surface supported by a structural element. The latter accounts for the increase in probability of the lack of total surface loading as the size of the surface area increases.

Roof surfaces must also be designed for wind pressure, for which the magnitude and manner of application are specified by local building codes based on local wind histories. For very light roof construction, a critical problem is sometimes that of the upward (suction) effect of the wind, which may exceed the dead load and result in a net upward lifting force.

Although the term *flat roof* is often used, there is generally no such thing; all roofs must be designed for some water drainage. The minimum required pitch is usually $\frac{1}{4}$ in./ft, or a slope of approximately 1 : 50. With roof surfaces that are this close to flat, a potential problem is that of *ponding,* a phenomenon in which the weight of water on the surface causes deflection of the supporting structure, which in turn allows for more water accumulation (in a pond) causing more deflection, and so on, resulting in an accelerated collapse condition.

20-4 Floor Live Loads

The live load on a floor represents the probable effects created by the occupancy. It includes the weights of human occupants, furniture, equipment, stored materials, and so on. All building codes provide minimum live loads to be used in the design of buildings for various occupancies. Since there is a lack of uniformity among different codes in specifying live loads, the local code should always be used. Table 20-3 contains values for floor live loads as given by the 1985 edition of the *Uniform Building Code*.

Although expressed as uniform loads, code-required values are usually established large enough to account for ordinary concentrations that occur. For offices, parking garages, and some other occupancies, codes often require the consideration of a specified concentrated load as well as the distributed loading. Where buildings are to contain heavy machinery, stored materials, or other contents of unusual weight, these must be provided for individually in the design of the structure.

When structural framing members support large areas, most codes allow some reduction in the total live load to be used for design. These reductions, in the case of roof loads, are incorporated into the data in Table 20-2. The following is the method given in the 1985 edition of the *Uniform Building Code* for determining the reduction permitted for beams, trusses, or columns that support large floor areas.

Except for floors in places of assembly (theaters, etc.), and except for live loads greater than 100 psf [4.79 kN/m^2], the design live load on a member may be reduced in accordance with the formula

$$R = 0.08\,(A - 150)$$

$$[R = 0.86\,(A - 14)]$$

The reduction shall not exceed 40% for horizontal members or for vertical members receiving load from one level only, 60% for other vertical members, nor R as determined by the formula

$$R = 23.1\left(1 + \frac{D}{L}\right)$$

TABLE 20-1. Weights of Building Construction

	lb/ft²	kN/m²
Roofs		
3-ply ready roofing (roll, composition)	1	0.05
3-ply felt and gravel	5.5	0.26
5-ply felt and gravel	6.5	0.31
Shingles		
wood	2	0.10
asphalt	2–3	0.10–0.15
clay tile	9–12	0.43–0.58
concrete tile	8–12	0.38–0.58
slate, 1/4 in.	10	0.48
fiber glass	2–3	0.10–0.15
aluminum	1	0.05
steel	2	0.10
Insulation		
fiber glass batts	0.5	0.025
rigid foam plastic	1.5	0.075
foamed concrete, mineral aggregate	2.5/in.	0.0047/mm
Wood rafters		
2 × 6 at 24 in.	1.0	0.05
2 × 8 at 24 in.	1.4	0.07
2 × 10 at 24 in.	1.7	0.08
2 × 12 at 24 in.	2.1	0.10
Steel deck, painted		
22 ga	1.6	0.08
20 ga	2.0	0.10
18 ga	2.6	0.13
Skylight		
glass with steel frame	6–10	0.29–0.48
plastic with aluminum frame	3–6	0.15–0.29
Plywood or softwood board sheathing	3.0/in.	0.0057/mm
Ceilings		
Suspended steel channels	1	0.05
Lath		
steel mesh	0.5	0.025
gypsum board, 1/2 in.	2	0.10
Fiber tile	1	0.05
Dry wall, gypsum board, 1/2 in.	2.5	0.12
Plaster		
gypsum, acoustic	5	0.24
cement	8.5	0.41
Suspended lighting and air distribution systems, average	3	0.15

Floors		
Hardwood, 1/2 in.	2.5	0.12
Vinyl tile, 1/8 in.	1.5	0.07
Asphalt mastic	12/in.	0.023/mm
Ceramic tile		
3/4 in.	10	0.48
thin set	5	0.24
Fiberboard underlay, 5/8 in.	3	0.15
Carpet and pad, average	3	0.15
Timber deck	2.5/in.	0.0047/mm
Steel deck, stone concrete fill, average	35–40	1.68–1.92
Concrete deck, stone aggregate	12.5/in.	0.024/mm
Wood joists		
2 × 8 at 16 in.	2.1	0.10
2 × 10 at 16 in.	2.6	0.13
2 × 12 at 16 in.	3.2	0.16
Lightweight concrete fill	8.0/in.	0.015/mm
Walls		
2 × 4 studs at 16 in., average	2	0.10
Steel studs at 16 in., average	4	0.20
Lath, plaster; see Ceilings		
Gypsum dry wall, 5/8 in. single	2.5	0.12
Stucco, 7/8 in., on wire and paper or felt	10	0.48
Windows, average, glazing + frame		
small pane, single glazing, wood or metal frame	5	0.24
large pane, single glazing, wood or metal frame	8	0.38
increase for double glazing	2–3	0.10–0.15
curtain walls, manufactured units	10–15	0.48–0.72
Brick veneer		
4 in., mortar joints	40	1.92
1/2 in., mastic	10	0.48
Concrete block		
lightweight, unreinforced—4 in.	20	0.96
6 in.	25	1.20
8 in.	30	1.44
heavy, reinforced, grouted—6 in.	45	2.15
8 in.	60	2.87
12 in.	85	4.07

TABLE 20-2. Minimum Roof Live Loads

	Minimum Uniformly Distributed Load (lb/ft²)			Minimum Uniformly Distributed Load (kN/m²)		
	Tributary Loaded Area for Structural Member (ft²)			Tributary Loaded Area for Structural Member (m²)		
Roof Slope Conditions	0–200	201–600	Over 600	0–18.6	18.7–55.7	Over 55.7
1. Flat or rise less than 4 in./ft (1 : 3). Arch or dome with rise less than 1/8 span.	20	16	12	0.96	0.77	0.575
2. Rise 4 in./ft (1 : 3) to less than 12 in./ft (1 : 1). Arch or dome with rise 1/8 of span to less than 3/8 of span.	16	14	12	0.77	0.67	0.575
3. Rise 12 in./ft (1 : 1) or greater. Arch or dome with rise 3/8 of span or greater.	12	12	12	0.575	0.575	0.575
4. Awnings, except cloth covered.	5	5	5	0.24	0.24	0.24
5. Greenhouses, lath houses, and agricultural buildings.	10	10	10	0.48	0.48	0.48

Source: Adapted from the *Uniform Building Code*, 1985 edition, with permission of the publishers, International Conference of Building Officials.

TABLE 20-3. Minimum Floor Live Loads

Use or Occupancy		Uniform Load		Concentrated Load	
Description	Description	(psf)	(kN/m²)	(lb)	(kN)
Armories		150	7.2		
Assembly areas and auditoriums and balconies therewith	Fixed seating areas	50	2.4		
	Movable seating and other areas	100	4.8		
	Stages and enclosed platforms	125	6.0		
Cornices, marquees, and residential balconies		60	2.9		
Exit facilities		100	4.8		
Garages	General storage, repair	100	4.8	*	
	Private pleasure car	50	2.4	*	
Hospitals	Wards and rooms	40	1.9	1000	4.5
Libraries	Reading rooms	60	2.9	1000	4.5
	Stack rooms	125	6.0	1500	6.7
Manufacturing	Light	75	3.6	2000	9.0
	Heavy	125	6.0	3000	13.3
Offices		50	2.4	2000	9.0
Printing plants	Press rooms	150	7.2	2500	11.1
	Composing rooms	100	4.8	2000	9.0
Residential		40	1.9		
Rest rooms		**			
Reviewing stands, grandstands, and bleachers		100	4.8		
Roof decks (occupied)	Same as area served				
Schools	Classrooms	40	1.9	1000	4.5
Sidewalks and driveways	Public access	250	12.0	*	
Storage	Light	125	6.0		
	Heavy	250	12.0		
Stores	Retail	75	3.6	2000	9.0
	Wholesale	100	4.8	3000	13.3

* Wheel loads related to size of vehicles that have access to the area.
** Same as the area served or minimum of 50 psf.

Source: Adapted from the *Uniform Building Code,* 1985 edition, with permission of the publishers, International Conference of Building Officials.

In these formulas

R = reduction in percent,
A = area of floor supported by a member,
D = unit dead load/sq ft of supported area,
L = unit live load/sq ft of supported area.

In office buildings and certain other building types, partitions may not be permanently fixed in location but may be erected or moved from one position to another in accordance with the requirements of the occupants. In order to provide for this flexibility, it is customary to require an allowance of 15 to 20 psf [0.72 to 0.96 kN/m^2] which is usually added to other dead loads.

20-5 Lateral Loads

Design for the effects of wind and earthquakes varies considerably on a regional basis and is typically controlled by the requirements of local building codes. The location of a building site will ordinarily place the building in the jurisdiction of a political body (city, county, state, or federal government) with powers to enforce a building code. Although techniques of design and the selection of specific systems and details of construction are subject to judgment of designers, the legal requirements of the building code of jurisdiction must be met.

Chapter 19 contains some discussion of concerns for lateral forces, with specific application to the design of diaphragm-braced, wood-framed structures. For a more complete treatment of the general subject of lateral force effects and design requirements the reader is referred to *Design for Lateral Forces* (Ref. 8). The work shown in the examples in this chapter is based on the discussions and illustrations in Chapter 19 in this book and generally conforms with the requirements of the 1985 edition of the *Uniform Building Code* (Ref. 3).

20-6 Building One: General Construction

Figure 20-1 shows a one-story, box-shaped building intended for commercial occupancy. Assuming that light wood framing is ade-

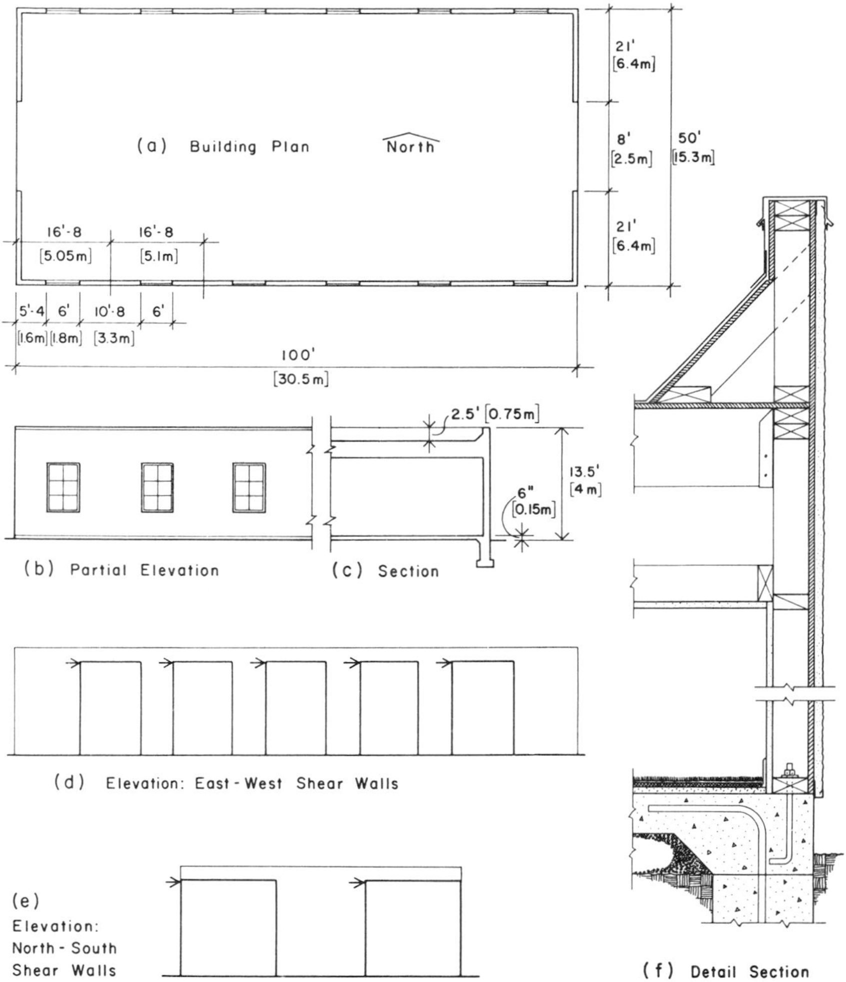

FIGURE 20-1. Building One: general form.

quate for fire resistance requirements, we will illustrate the design of the major elements of the structure for this building. We assume the following data for design:

Roof live load = 20 psf (reducible).
Wind load critical at 20 psf on vertical exterior surface.
Wood framing of Douglas fir–larch.

The general profile of the building is indicated in Fig. 20-1*c*, which shows a flat roof and ceiling and a short parapet at the exterior walls. The general nature of the construction is shown in the detailed wall section in Fig. 20-1*f*. Specific details of the framing will depend on various decisions in the design of the structure as developed in the following discussion. The general form of the exterior shear walls is indicated in Figs. 20-1*d* and *e*. Design considerations for lateral loads are presented in Sec. 20-8. We first consider the design of the roof structural system for only gravity loads, although we keep in mind that the roof must eventually be developed as a horizontal diaphragm and the walls as shear walls.

20-7 Building One: Design for Gravity Loads

With the construction as shown in Fig. 20-1*f*, we determine the roof loads as follows:

Three-ply felt and gravel roofing	5.5 psf
Fiberglass insulation batts	0.5
$\frac{1}{2}$ in. thick plywood roof deck	1.5
Rafters and blocking (estimate)	2.0
Ceiling joists	1.0
$\frac{1}{2}$ in. drywall ceiling	2.5
Ducts, lights, etc.	3.0
Total roof dead load	16.0 psf

Assuming a partitioning of the interior as shown in Fig. 20-2*a*, various possibilities exist for the development of the spanning roof and ceiling framing systems and their supports. Interior walls

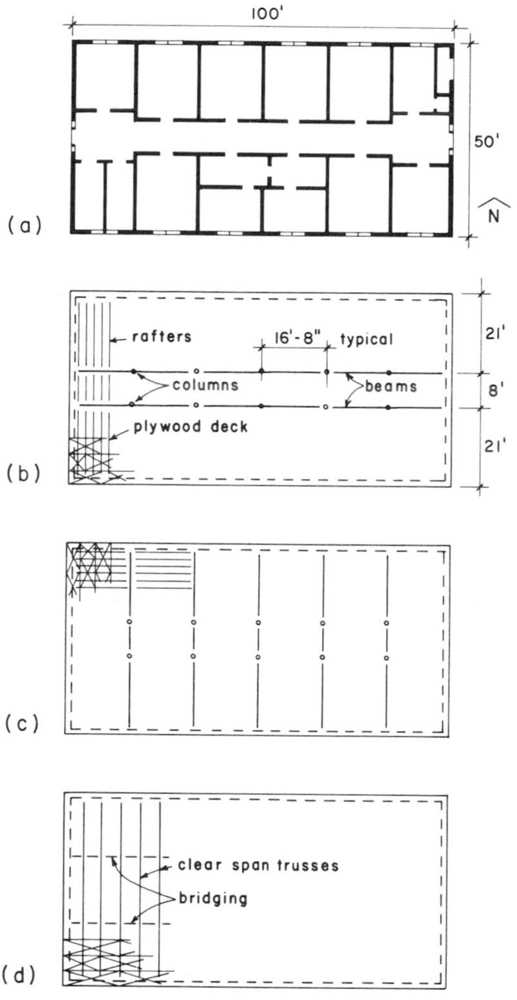

FIGURE 20-2. Alternatives for the roof framing.

may be used for supports, but a more desirable situation in commercial uses is sometimes obtained by using interior columns that allow for rearrangement of the interior spaces. The roof-framing system shown in Fig. 20-2*b* is developed with two rows of interior columns placed at the location of the corridor walls. If the partitioning shown in Fig. 20-2*a* is used, these columns will be architecturally out of view because they are incorporated within the wall space.

Figure 20-2*c* shows a second possibility for the roof framing using the same column layout as in Fig. 20-2*b*. There may be various reasons for favoring one of these framing schemes over the other. Problems of installation of ducts, lighting, wiring, roof drains, and fire sprinklers may influence the choice. We arbitrarily select the scheme shown in Fig. 20-2*b* to illustrate the design process for the elements of the structure.

Installation of a membrane-type roofing ordinarily requires at least a $\frac{1}{2}$ in. thick roof deck. Such a deck is capable of up to 32 in. spans in a direction parallel to the face ply grain (the long dimension of ordinary 4 ft by 8 ft panels). If rafters are not over 24 in. on center—as they are likely to be for the schemes shown in Figs. 20-2*b* and *c*—the panels may be placed so that the plywood span is across the face grain. An advantage in the latter arrangement is the reduction in the amount of blocking required at edges not falling on the rafter. The reader is referred to the general discussions in Chapter 17 for uses of plywood structural decking.

For the rafters we assume No. 2 grade, an ordinary minimum structural usage. From Table 3-1 the allowable bending stress for repetitive use is 1450 psi and the modulus of elasticity is 1,700,000 psi. Since our loading falls approximately within the criteria of Table 10-3, we may use that table as a reference. Inspection of the table yields the possibilities of using either 2 × 10s at 16 in. centers or 2 × 12s at 24 in. centers. The reader can verify the adequacy of either of these choices using the procedures developed in Chapter 9. For stress investigation it is likely that an increase of the allowable stress of either 15% or 25% would be permitted. (See discussion of adjustment for load duration in Chapter 3.)

If the ceiling joists also span the full 21 ft, an inspection of Table 10-2 yields the possibilities of using either 2 × 8s at 16 in. centers or 2 × 10s at 24 in. centers. Spacing of ceiling joists must, of course, relate to the surfacing material. It is also possible to either suspend the ceiling joists from the rafters or to use the interior partitions for support. The dead load tabulation made previously assumes the use of light ceiling framing—most likely 2 × 4 or 2 × 3 members—suspended from the rafters.

The beams as shown in Fig. 20-2*b* are continuous through two spans, with a total length of 33 ft 4 in. and a clear span of 16 ft 8 in. For the two-span beam the maximum bending moment is the same as that for a simple span, the principal advantage being a reduction of deflection. The total load area for one span is

$$16.67 \times \frac{21 + 8}{2} = 242 \text{ ft}^2$$

As indicated in Table 20-2, this permits the use of a live load of 16 psf. Thus the unit of the uniformly distributed load on the beam is found as

$$(16 + 16) \times \left(\frac{21 + 8}{2}\right) = 464 \text{ lb/ft}$$

Adding a bit for the beam weight, we will design for a load of 480 lb/ft. The maximum moment is thus found to be

$$M = \frac{wL^2}{8} = \frac{480(16.67)^2}{8} = 16{,}673 \text{ ft-lb}$$

The allowable bending stress depends on the beam size and the load duration assumed. We will assume a 15% increase for load duration and thus determine the following (see Table 3-1):

For a 4× member: $F_b = 1.15(1500) = 1725$ psi

For a 5× member or larger: $F_b = 1.15(1300) = 1495$ psi

Then

$$S = \frac{M}{F_b} = \frac{16{,}673 \times 12}{1725} = 116 \text{ in.}^3$$

thus indicating a 4 × 16 as a possible choice ($S = 135.7$ in.^{3}). Or

$$S = \frac{16{,}673 \times 12}{1495} = 134 \text{ in.}^3$$

thus indicating a 6 × 14 ($S = 167$ in.^{3}) or an 8 × 12 ($S = 165$ in.^{3})

Although the 4 × 16 offers the least cross section area and ostensibly the lower cost, various considerations of the development of the construction details may affect the beam selection. The beam can also be formed as a built-up member from a number of 2× elements for which the allowable stress is even higher. The latter may be preferred where good quality heavy timber sections are not easily obtained.

Finally, the beam may consist of a glue-laminated or rolled steel section. Advantages of these choices are a reduction in beam depth and a somewhat better long-term deflection response. This is likely to be more critical for beams of longer span, for example, the beams in the framing scheme in Fig. 20-2*c*.

The beam should also be investigated for shear and deflection. Note that the maximum shear in the two-span beam is slightly larger than the simple beam shear of $wL^2/2$. This may be a reason for *not* using the two-span condition in wood beams where shear is often a critical concern for solid-sawn sections. For the example here, the span is quite short so that deflection is not a critical concern, especially if the two-span condition is used. Shear is also not critical if the stress increase for load duration is made and the load on the ends of the spans is eliminated, as discussed in Chapter 6.

We assume that a minimum roof slope of $\frac{1}{4}$ in. per ft is required to drain the roof surface. If the draining of the roof occurs at the outside walls, the building profile may thus be as shown in Fig. 20-3*a*. This requires a total elevation change of approximately $\frac{1}{4} \times 25 = 6.25$ in. from the center to the long side of the building.

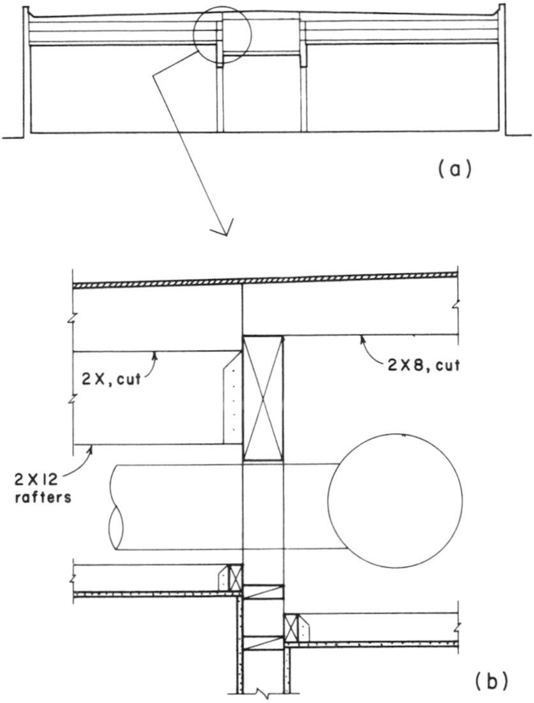

FIGURE 20-3. Building One: construction details.

There are various ways to achieve this, including the tilting of the rafters.

Figure 20-3*b* shows some possibilities for the details of the construction at the center of the building. In this view the rafters are kept flat and the roof profile for drainage is achieved by attaching cut 2× elements on top of the rafters and using short profiled rafters at the corridor. Ceiling joists for the corridor are supported directly by the corridor walls. Other ceiling joists are supported at their ends by the walls, although long span ceiling joists could be suspended in their spans by hangers from the rafters. Ceiling construction may be developed together with the roof structure or may be developed with the interior partitioning, depending on the occupancy needs.

The typical interior column supports a load approximately equal to the entire load on one beam or

$$P = 480 \times 16.67 = 8000 \text{ lb}$$

This is quite a light load, but the column height requires larger than a 4× size. (See Table 12-1.) If a 6 × 6 is not objectionable, it is adequate in the lower stress grades. However, it may be better to consider use of a steel pipe or tubular section, either of which can be accommodated in a partition wall with 2 × 4 studs.

Structural design of the studs in the exterior walls is principally a matter of consideration for lateral bending due to wind. This is considered in the next section.

20-8 Building One: Design for Wind

Design of the building structure for wind includes the consideration for the following:

1. Inward and outward pressure on exterior walls, causing bending of the wall studs.
2. Total lateral force on the building, requiring bracing by the roof diaphragm and the shear walls.
3. Uplift on the roof, possibly requiring anchorage of the roof structure.
4. Total effect of lateral and uplift forces, possibly resulting in overturn of the entire building.

We will first investigate the wall studs on the north and south sides (see Fig. 20-1*a*), assuming them to be 2 × 6 members of Douglas fir–larch, Stud grade, and to be on 16 in. centers. These walls support the ends of the 21 ft span rafters as well as the parapet wall above. We will assume the total gravity dead load plus live load on the wall to be approximately 400 lb/ft of wall length. Thus a single stud supports a load of

$$P = 400 \times \frac{16}{12} = 533 \text{ lb}$$

Assuming the studs to span 10.5 ft vertically, each stud will carry a uniformly distributed wind load of $w = 20$ psf $\times$ (16/12) = 26.7 lb/ft and will thus sustain a maximum bending moment of

$$M = \frac{wL^2}{8} = \frac{26.7(10.5)^2}{8} = 368 \text{ ft-lb}$$

which results in a maximum bending stress in the stud of

$$f_b = \frac{M}{S} = \frac{368 \times 12}{7.563} = 584 \text{ psi}$$

The axial compression stress on a single stud is

$$f_c = \frac{P}{A} = \frac{533}{8.25} = 65 \text{ psi}$$

For Stud grade wood, Table 3-1 yields the following: $F_b = 850$ psi (repetitive member use); $F_c = 675$ psi; $E = 1{,}500{,}000$ psi. For the allowable column compression stress (F'_c) the slenderness ratio is determined as $L/d = (10.5 \times 12)/5.5 = 22.9$. To establish the slenderness category, we determine

$$K = 0.671\sqrt{\frac{E}{F_c}} = 0.671\sqrt{\frac{1{,}500{,}000}{675}} = 31.6$$

Thus the allowable compression is determined as

$$F'_c = F_c\left[1 - \frac{1}{3}\left(\frac{L/d}{K}\right)^4\right] = 675\left[1 - \frac{1}{3}\left(\frac{22.9}{31.6}\right)^4\right] = 613 \text{ psi}$$

Then the combined axial compression and lateral bending condition is investigated as follows:

$$\frac{f_c}{F'_c} + \frac{f_b}{F'_b - Jf_c} = \frac{65}{1.33(613)} + \frac{584}{1.33(850 - 65)}$$

$$= 0.080 + 0.560 = 0.640$$

As this total is less than one, the 2 × 6 stud is adequate in the Stud grade. An alternate solution would be to use a 2 × 4 in a higher stress grade such as No. 1. However, economy will probably be achieved by using the Stud grade if 2 × 6 size studs can be obtained in this grade.

Consideration for uplift force on the roof varies with different building codes. The maximum consideration usually consists of an uplift equal to the horizontal design pressure. If this is required, the uplift force exceeds the roof dead load by 4 psf in this example, and the roof must be anchored to its supports to resist this effect. With the use of metal-framing anchors between the rafters and the beams or between the rafters and the stud walls, adequate anchorage will probably be provided.

Overturning of the entire building is not likely to be critical for a building with the profile in this example. Overturn is more critical in buildings of extremely light dead weight or those that are quite tall and narrow in profile. In any event, the investigation is similar to that for overturn of a single shear wall except for the inclusion of the uplift effect on the roof. If the total overturning moment is more than two-thirds of the available resisting moment due to building dead load, the anchorage of the building as a whole must be considered.

We proceed to consideration of the forces exerted on the principal bracing elements, that is, the roof diaphragm and the exterior shear walls. The building must be investigated for wind in the two directions—east–west and north–south. Consideration of the wall functions and determination of the forces exerted on the bracing system are illustrated in Fig. 20-4. The amount of the portion of the pressure on the exterior walls that is applied as load to the edge of the roof diaphragm depends partly on the spanning nature of the exterior walls. Figure 20-4*a* shows two common cases: with studs cantilevering to form the parapet and with simple span studs and a separate parapet structure. With the construction as shown in Fig. 20-1*f*, we assume the walls on the north and south sides to be Case 2 in Fig. 20-4*a*. Thus the lateral wind load applied to the roof diaphragm in the north–south direction is

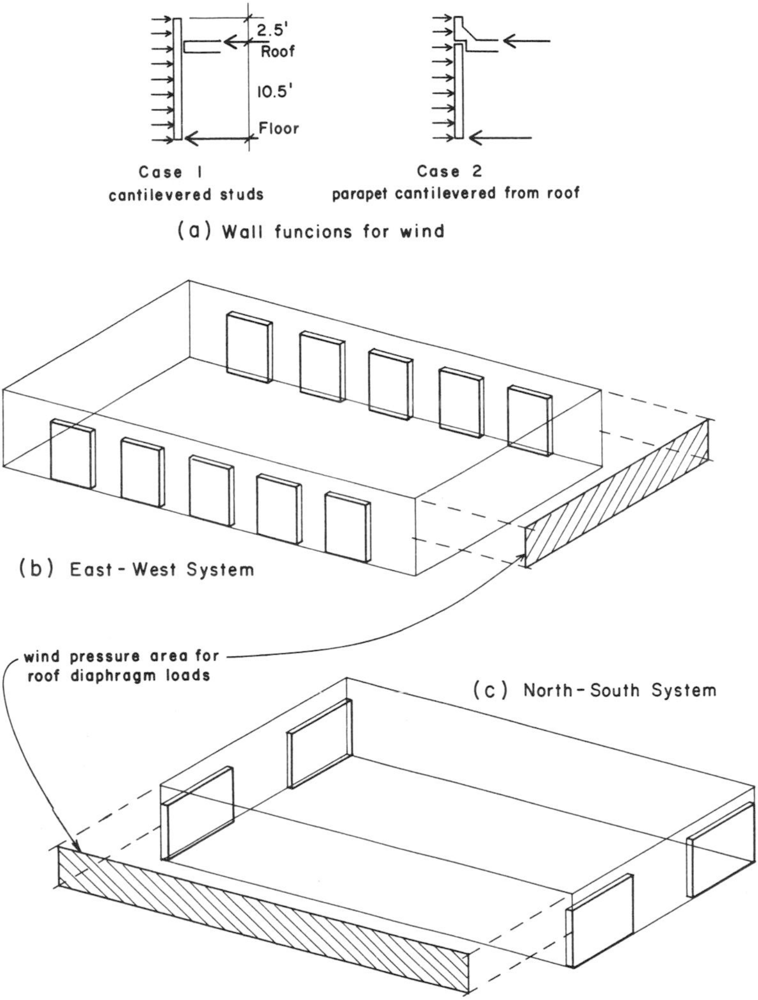

FIGURE 20-4. Building One: wall functions and wind pressure development.

$$(20 \text{ psf})\left(\frac{10.5}{2}\right) + (20 \text{ psf})(2.5) = 155 \text{ lb/ft}$$

In resisting this load, the roof functions as a spanning member supported by the shear walls at the east and west ends of the building. The investigation of this 100 ft span simple beam with uniformly distributed loading is shown in Fig. 20-5. The end reaction and maximum shear force is found as

$$(155)\left(\frac{100}{2}\right) = 7750 \text{ lb}$$

which results in a maximum unit shear in the 50 ft wide roof diaphragm of

$$v = \frac{\text{shear force}}{\text{roof width}} = \frac{7750}{50} = 155 \text{ lb/ft}$$

From Table 19-1 (*UBC,* Table 25-J-1) we may select a number of possible choices for the roof deck. Variables include the class of the plywood, the panel thickness, the width of supporting rafters, the nail size and spacing, and the use or omission of blocking (support for nailing at the plywood panel edges not falling on the rafters). In addition to satisfying the shear capacity requirement, there are many considerations for the choice of the construction that derive from the design for gravity loading, general construction details for the roof structure, and the problems of installation of roofing and insulation. For the flat roof with ordinary tar and felt membrane roofing, it is usually necessary to provide a minimum of $\frac{1}{2}$ in. (now $\frac{15}{32}$ in.) thick plywood. If this requirement is accepted, a possible choice from Table 19-1 is

> Structural II 15/32 in. plywood with 2× framing and 8d nails at 6 in. at all panel edges and a blocked diaphragm.

For these criteria Table 19-1 yields a capacity of 270 lb/ft.

In this example, if the need for the minimum thickness of plywood is accepted, it turns out that the minimal construction is

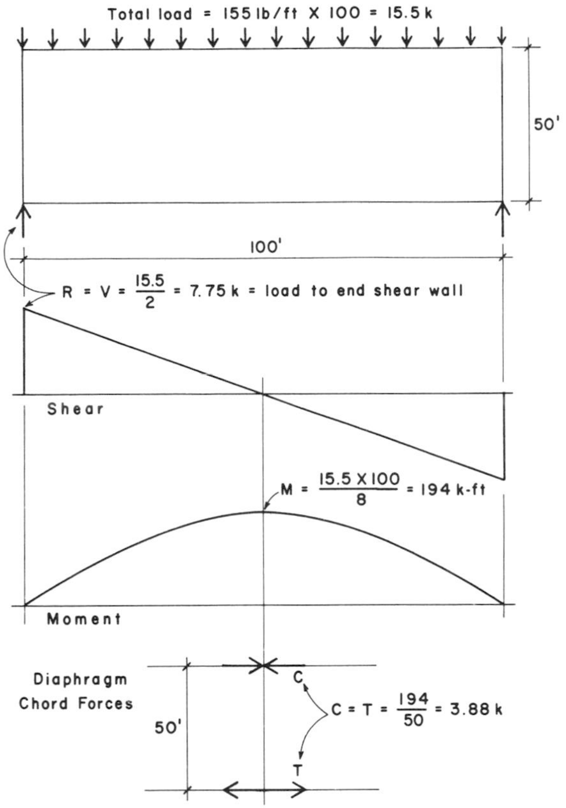

FIGURE 20-5. Functions of the roof diaphragm.

more than adequate for the required lateral force resistance. Had this not been the case, and the required capacity had resulted in considerable nailing beyond the minimal, it would be possible to graduate the nail spacings from the maximum required at the building ends to minimal nailing in the center portion of the roof. (See Example 3 in Sec. 19-2.)

The moment diagram shown in Fig. 20-5 indicates a maximum value of 194 k-ft at the center of the span. This moment is used to determine the required chord force that must be developed in both tension and compression at the roof edges. With the construction as shown in Fig. 20-1*f*, the top plate of the stud wall is the most likely element to be used for this function. In this example the force is quite small and can be easily developed by the ordinary construction. The one problem that may require some special effort is the development of the member as a continuous tension element, since it is not possible to have a single, 100 ft long piece for the plate. Splicing of the multipiece plates must therefore be developed to provide a continuity of the tension force.

The end reaction force for the roof diaphragm, as shown in Fig. 20-5, must be developed by the end shear walls. As shown in Fig. 20-1, there are two walls at each end, both 21 ft long in plan. Thus the total shear force is developed by a total of 42 ft of shear wall and the unit shear in the walls is

$$v = \frac{\text{total shear force}}{\text{total wall length}} = \frac{7750}{42} = 185 \text{ lb/ft}$$

As with the roof deck, there are various considerations for the selection of the wall construction. A common situation involves the use of a single facing of structural plywood on the exterior surface of the wall, which is considered as the resisting element for lateral force. Other elements of the wall construction are thus considered nonstructural in function. For this situation a possible choice from Table 19-2 (*UBC*, Table 25-K-1) is

Structural II, $\frac{3}{8}$ in. thick plywood with 6d nails at 6 in. spacing at all panel edges.

Again, this is minimal construction. For situations that require much higher capacities of shear resistance it may be necessary to use thicker than normal plywood and nailing with larger, closer spaced nails. Unfortunately, the nailing cannot be graduated—as it may be for the roof—as the unit shear is a constant value throughout the height of the shear wall.

Figure 20-6*a* shows the loading condition for investigating the overturning effect on the shear wall. The loading shown includes

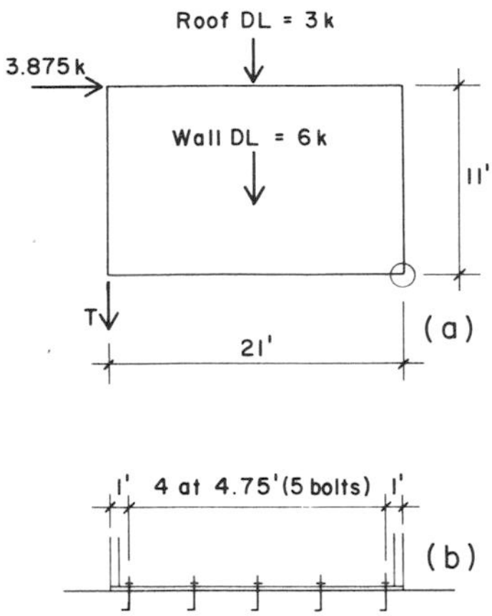

FIGURE 20-6.

only the lateral force (applied at the level of the roof deck) and the dead loads of the wall itself plus the portion of the roof carried by the wall. The overturning moment is determined as the product of the lateral force and its distance above the wall base. This value is then multiplied by a factor of 1.5, representing the usual minimum required safety factor for comparison with the dead load resisting moment (also called the restoring moment). If the minimum safety factor is not provided, an anchorage force (T in Fig. 20-6*a*) must be added to supplement the dead load resistance. The usual computations for this investigation are as follows:

$$\text{overturning moment} = (3.875)(11)(1.5) = 64 \text{ k-ft}$$

$$\text{restoring moment} = (3 + 6)(21/2) = 94.5 \text{ k-ft}$$

This indicates that no tiedown force is required. Actually, there is additional resistance at each end of the wall. At the building corner this wall is most likely quite well attached to the wall on the north or south side of the building, which provides additional dead load resistance. At the end of the wall near the corridor a post would be provided for support of the beam (see the framing plan in Fig. 20-2). Thus the dead load portion of the beam reaction provides additional resistance. Finally, the wall sill will be bolted to the foundation providing some hold down resistance to toppling of the wall. At present, most codes do not permit use of a computed value for the sill bolt resistance to uplift as this function involves cross-grain bending of the sill.

The sill bolts *will* be used for resistance to the horizontal sliding of the wall, however, and the bolting must satisfy this requirement. Code minimum bolting usually consists of $\frac{1}{2}$ in. diameter bolts at 1 ft from the wall ends and at a maximum of 6 ft on center for the remainder of the wall length. A layout for this minimal bolting is shown in Fig. 20-6*b*. With a 2× sill member and $\frac{1}{2}$ in. bolts in single shear, Table 13-2 yields a value of 470 lb for one bolt. With a total of five bolts and an increase of one-third in the value for wind loading, the total sliding resistance of the minimum bolting is

$$(1.33)(470)(5) = 3125 \text{ lb}$$

As this is a bit short of the necessary resistance, it is necessary to increase the number and/or the size of the bolts. Within the limit of the sill width, it is probably best to increase the bolt size for economy. Setting of the bolts in the concrete is a major cost factor that is not much related to the bolt size. Thus a few large bolts are probably preferable to a lot of small bolts.

In some situations it may be necessary to consider the effects of the sliding and overturn forces on the wall foundations. In this example the forces are quite low and not likely to be critical. For buildings with shallow foundations, however, the overturning and sliding must be seriously considered in the foundation design when they are high in magnitude in relation to gravity forces.

A major consideration in the design for lateral force resistance is the development of the necessary force transfers between the various elements of the lateral bracing system. The sill bolting of the shear wall is one example of such a transfer. Another critical one is the transfer of force from the roof diaphragm to the shear wall. In this example the roof shear is vested in the roof plywood and the wall shear in the wall plywood. As these two elements are not directly attached, the details of the framing must be studied to determine how the transfer can be made. With the construction as shown in Fig. 20-1*f* a simple transfer can be made through the top plate of the wall to which both the roof and wall plywoods are directly attached. This is likely to occur normally for the roof plywood because this is the edge of the roof diaphragm (called the *boundary* in Table 19-1). For the wall plywood, however, this may not be a panel edge, and the necessary nailing for the required transfer should be indicated and specified on the wall section.

For other wall and roof constructions the force transfer may not be as simple and direct as that in Fig. 20-1*f*. In some cases there may not be any direct transfer through ordinary elements of the construction, thus requiring the addition of some framing or connecting devices.

20-9 Building Two: General Construction

Figure 20-7 shows a building that consists essentially of stacking the plan for Building One to produce a two-story building. The profile section of the building shows that the structure for the second floor is developed essentially the same as the roof structure for Building One. For the roof, however, a clear span structure could be provided by 50 ft span trusses—a relatively easy task for prefabricated, manufactured trusses as described in Chapter 15. This roof structure would also be an alternative for Building One and would provide for a greater degree of freedom for the arrangement of the partitioning in the building interior.

Although the second floor structure for this building is similar to the roof structure for Building One, the principal difference is in the magnitude of live load that must be accommodated. The usual building code requirement for minimum load for offices is 50 psf (see Table 20-3). However, it is also usually required to add an allowance of 20 or 25 psf for movable partitioning. If we use a total live load of 75 psf, the design load will be several times greater than that for the roof. The difference in results is demonstrated in the next section.

The two-story building will also sustain a greater total wind force, although the shear walls in the second story will essentially

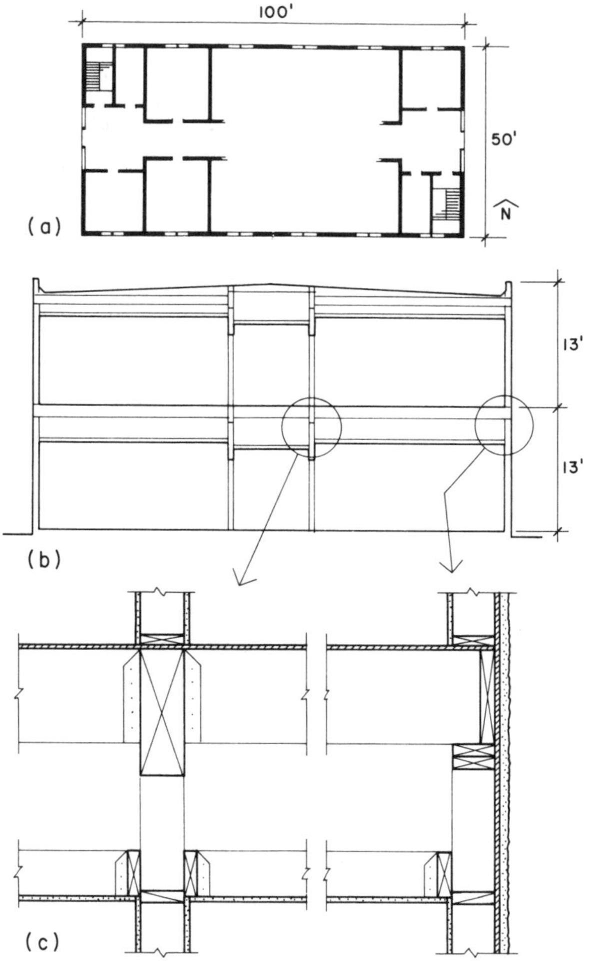

FIGURE 20-7. Building Two: general form.

be the same as those for Building One. The major effect in this building will be the forces in the first-story shear walls.

Another difference in the floor structure is that it will be dead flat, thereby eliminating the concerns for producing the sloped surface for draining of the roof. The details for the second floor framing are shown in Fig. 20-7*c*, indicating the use of platform framing. Otherwise the details of the construction will be assumed to be similar to those for Building One except for the possible use of the trusses for the roof. One way to economize with the trusses is by eliminating a general framing for the ceiling, and having the ceiling surface attached directly to the bottom chords of the trusses. Another economy is possible if the top chords of the trusses are sloped to provide the framing for the roof deck in the profile necessary for drainage. The planning of the building, the location of roof drains, and the framing plan for the trusses must be coordinated to bring these simplifications off effectively.

20-10 Building Two: Design of the Floor Structure

Because of the increased loads, the member sizes for the framing of the floor with solid-sawn lumber will be considerably greater than those required for the roof for Building One. We now illustrate the design with structural lumber, but will discuss alternatives later.

We assume the following for the dead load of the floor construction, assuming the ceiling to be separately supported (not hung from the floor joists):

Carpet and pad	3.0 psf
Fiberboard underlay	3.0
Plywood deck, $\frac{1}{2}$ in.	1.5
Ducts, lights, wiring	3.5
Total without joists:	11.0 psf

With joists at 16 in. spacing, the loading for a single joist is

$$DL = \frac{16}{12}(11) = 14.6 \text{ lb/ft} + \text{the joist, say } 20 \text{ lb/ft}$$

$$LL = \frac{16}{12}(75) = 100 \text{ lb/ft}, \quad \text{total} = 120 \text{ lb/ft}$$

For the 21 ft span joists the maximum bending moment is

$$M = \frac{wL^2}{8} = \frac{(120)(21)^2}{8} = 6615 \text{ ft-lb}$$

For No. 1 Douglas fir–larch joists with 2 in. nominal thickness and 5 in. width or wider, F_b from Table 3-1 is 1750 psi for repetitive member use. Thus the required section modulus is

$$S = \frac{M}{S} = \frac{(6615)(12)}{1750} = 45.36 \text{ in.}^3$$

From Table 4-1 we find that this is just over the value for a 2 × 14, which is the largest member listed with a 2 in. nominal thickness. Alternatives are to increase the stress grade of the wood, use a thicker joist, or reduce the joist spacing. If the spacing is reduced to 12 in., the required section modulus drops by almost one-fourth, making the 2 × 14 adequate for flexure. Shear is unlikely to be critical with long span joists, but deflection should be investigated.

The usual deflection limit for this situation is a maximum live load deflection of 1/360 of the span, or (21)(12)/360 = 0.7 in. With the 2 × 14 at 12 in. spacing, the maximum deflection under live load only is

$$\Delta = \frac{5}{384}\frac{wL^4}{\text{EI}} \quad \text{or} \quad \frac{5}{384}\frac{WL^3}{\text{EI}}$$

$$= \frac{5}{384}\frac{(75 \times 21)(21 \times 12)^3}{(1{,}800{,}000)(291)} = 0.63 \text{ in.}$$

which indicates that deflection is not critical.

The beams support both the 21 ft joists and the shorter 8 ft joists at the corridor. The corridor live load is usually required to be 100 psf, but the short span will probably permit a small joist—usually a 2 × 6 minimum. Thus the dead load at the corridor is slightly less. Also the total area supported by a beam exceeds 150 sq ft, which allows for some minor reduction of the live load as described in Sec. 20-4. We will simplify the computations by using the dead load and the live load as determined for the 21 ft span joists as the design load for the beam for the total of 14.5 ft of joist span (half the distance to the other beam plus half the distance to the exterior wall). Thus the beam load is

$$DL = (16)(14.5) = 232 \text{ lb/ft} \quad \text{(joist load)}$$
$$+ \text{ beam weight} = 30 \quad \text{(assumed estimate)}$$
$$+ \text{ wall above} = \underline{150} \quad \text{(2nd floor corridor)}$$
$$\text{Total } DL = 412 \text{ lb/ft}$$
$$LL = (75)(14.5) = 1088 \text{ lb/ft}$$
$$\text{Total load} = 412 + 1088 = 1500 \text{ lb/ft}$$

For a uniformly loaded simple span beam with a span of 16 ft 8 in. we determine

Total load $= W = (1.5)(16.67) = 25$ kips.
End reaction and maximum shear $= W/2 = 12.5$ kips.
Maximum moment $= wL^2/8 = (1.5)(16.67)^2/8 = 52.1$ k-ft.

For a Douglas fir–larch, Dense No. 1 grade beam, Table 3-1 yields values of $F_b = 1550$ psi, $F_v = 85$ psi, and $E = 1{,}700{,}000$ psi. To satisfy the flexural requirement, the required section modulus is

$$S = \frac{M}{F_b} = \frac{(52.1)(12)}{1.550} = 403 \text{ in.}^3$$

From Table 4-1 the least weight section that will satisfy this requirement is an 8 × 20 with $S = 475$ in.^{3}.

If the 20 in. deep section is used, its effective bending resistance must be reduced, as discussed in Sec. 7-4. Thus the actual moment capacity of this section is reduced by use of the size factor from Table 7-1 and is determined as

$$M = C_F \times F_b \times S = (0.947)(1.550)(475)(1/12) = 58.1 \text{ k-ft}$$

As this exceeds the requirement, the correction for size effect is not critical in the choice of the section.

If the actual beam depth is 19.5 in., the critical shear force may be reduced to that at a distance of the beam depth from the support. Thus we may subtract an amount of the load equal to the beam depth times the unit load. The critical shear force is thus

$$V = (\text{actual end shear}) - (\text{unit load times beam depth})$$

$$= 12.5 \text{ kips} - (1.5)\left(\frac{19.5}{12}\right) = 12.5 - 2.44 = 10.06 \text{ kips}$$

If the 8 × 20 is selected, the maximum shear stress is thus

$$f_v = \frac{3}{2}\frac{V}{A} = \frac{(3)(10{,}060)}{(2)(146.25)} = 103 \text{ psi}$$

Even with the reduction in the critical shear force, this exceeds the allowable stress of 85 psi. Thus the beam section must be increased to a 10 × 20 to satisfy the shear requirement. With this section, the flexural stress will be reduced and the use of the dense grade of wood may be unnecessary because the shear stress is a constant for all grades of the wood.

For beams of relatively short span and heavy loading, it is common for shear to be a controlling factor. This often rules against the practicality of using a solid-sawn timber section for which allowable shear stress is quite low. It is probably logical to modify the structure to reduce the beam span or to chose a steel beam or a glue-laminated section in place of the solid timber.

Although deflection is often critical for long span, lightly loaded joists, it is seldom critical for the short span, heavily loaded beam. The reader should verify this by investigating the deflection of this beam, but we will dispense with the computation.

For the interior column at the first story the design load is approximately the same as the total load on a single beam, that is, 25 kips. For the approximately 10 ft high column, Table 12-1 indicates a 6 × 6 column for a solid-sawn section. For various reasons it may be more practical to use a steel round pipe or a square tubular section whose size can fit into a 2 × 4 stud wall.

Columns must also be provided at the ends of the beams in the east and west walls. In the example these locations are the ends of the shear walls, and the normal use of a doubled stud at this point will probably result in an adequate column. If this location occurs in the center portion of a wall, a separate column, or special doubled stud, should be provided.

20-11 Building Two: Design for Wind

The general design for wind includes the considerations enumerated at the beginning of Sec. 20-7 for Building One. Investigation of the second-story studs in the exterior walls would be similar to that made for Building One. At the first story in Building Two the studs carry considerably more axial compression, but the bending due to wind is approximately the same as at the second story. The 2 × 6 studs at 16 in. centers are probably adequate at the first story. If an investigation similar to that made in Sec. 20-7 shows an overstress condition, the stud spacing can be reduced to 12 in. or a higher grade wood can be used.

For lateral load the roof deck in Building Two is basically the same as that in Building One. With the trusses it may be more practical to use an unblocked deck, and the investigation should consider this factor. The footnotes to Table 19-1 (*UBC,* Table 25-J-1) should be studied with regard to the pattern of the layout of the plywood panels, especially for unblocked decks. Various special deck panels with tongue–and–groove edges are available in thicknesses greater than $\frac{1}{2}$ in., thus permitting truss spacings up to 4 ft. Special data for lateral load resistance may be available from the manufacturers for these products, although local code approval must be determined.

The wind loading condition for the two-story building is shown in Fig. 20-8*a*. This indicates a loading to the second floor diaphragm of 235 lb/ft. With a 15/32 in. thick deck as a minimum, the shear in the 50 ft wide deck will not be critical. However, the stair wells at the east and west ends reduce the actual diaphragm width at the ends to only 35 ft. Figure 20-8*b* shows the loading for the second floor deck and the critical shear and moment values for the diaphragm actions. At the ends the critical unit shear in the deck is

$$v = \frac{11{,}750}{35} = 336 \text{ lb/ft}$$

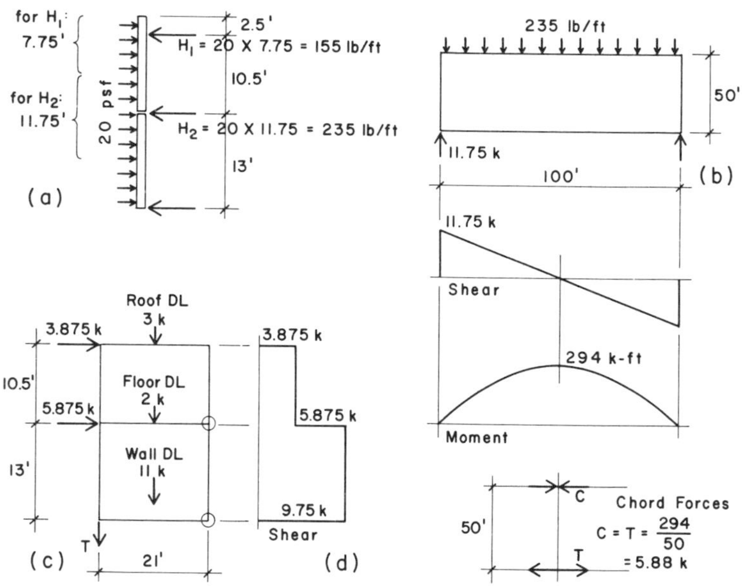

FIGURE 20-8. Building Two: development of lateral force: (*a*) generation of wind loads to the horizontal diaphragms; (*b*) functions of the second floor diaphragm; (*c*) loading of the end shear wall; (*d*) shear diagram for the two-story wall.

From Table 19-1 (*UBC*, Table 25-J-1) it may be determined that this requires a bit more than the minimum nailing for the deck. Options at this location include:

1. Using a 15/32 in. Structural II deck with 8d nails at 4 in. at the diaphragm boundary and other critical edges.
2. Using 15/32 in. Structural I deck with 10d nails at 6 in. throughout and 3× framing.
3. Using 19/32 in. Structural II deck with 10d nails at 6 in. throughout and 3× framing.

At 8 ft from the building ends the deck resumes its full 50 ft width, and the unit shear at this point drops to

$$v = \frac{9870}{50} = 197 \text{ lb/ft}$$

Since this value is well below the capacity of the 15/32 in. Structural II deck with minimum nailing, it may be the most practical to elect option (1), which involves only the use of 4 in. nailing in approximately 12% of the total second floor deck.

The diaphragm chord force for the second floor deck is approximately 6 kips, and must be developed in the framing at the wall, as shown in Fig. 20-7*c*. The most likely member to use for this is the continuous edge member at the face of the joists. The only real design consideration for this situation is developing the splicing of the member that will be made up of several pieces in the 100 ft length. Splicing may be achieved in a number of ways, the details must be developed to work within the construction of the wall and floor at this location. A joint using a steel strap with wood screws or one with bolts and steel plates will likely cause the least intrusion in the construction.

The loading for the two-story end shear wall is shown in Fig. 20-8*c* and the shear diagram for this load is shown in Fig. 20-8*d*. The second-story wall is essentially similar to the end wall in Building One for which an investigation was made in Sec. 20-8.

Because minimal construction is adequate here, and no anchorage for overturn is required, the only problem for concern is the development of sliding resistance to the lateral force of 3875 lb. Since this wall does not sit on a concrete foundation, other means of anchorage than steel anchor bolts must be considered.

The lateral force in the second-story wall must be transferred to the lower (first-story) wall. Essentially this occurs directly if the plywood is continuous past the construction at the level of the second floor, as it indeed is as shown in Fig. 20-7*c*. A critical location for stress transfer in this location is at the top of the second floor joists. At this point the lateral force from the second floor deck is transferred to the wall through the continuous edge member. Therefore the nailing and plywood requirements for the first-story wall begin at this location. The last point for the nailing and plywood requirements for the second-story wall are at the location of the sill for the second-story wall (on top of the second floor deck, as shown in Fig. 20-7*c*). The plywood for the wall and its nailing from this point down must satisfy the requirements for the first-story wall.

In the first-story wall the total shear force is 9750 lb and the unit shear is

$$v = \frac{9750}{21} = 464 \text{ lb/ft}$$

If the $\frac{3}{8}$ in. Structural II plywood selected for Building One is used for the second floor (see Sec. 20-7), it may be practical to use the same plywood for the entire two story wall and to simply increase the nail size and/or reduce the nail spacing at the first story. Table 19-2 (*UBC*, Table 25-K-1) yields a value of 410 lb/ft for $\frac{3}{8}$ in. Structural II plywood with 8d nails at 3 in. If the conditions of table footnote 3 are met, this value can be increased by 20% to 492 lb/ft. There are other options and many other design considerations for the choice of the wall construction, but this is an adequate choice for the lateral design criteria.

At the first floor level, the investigation for overturn of the end shear wall is as follows (see Fig. 20-8*c*):

$$\begin{aligned}
\text{overturning moment} &= (3.875)(23.5)(1.5) = 136.6 \text{ k-ft} \\
&\quad + (5.875)(13)(1.5) = \underline{114.6} \text{ k-ft} \\
&\qquad\qquad \text{Total} = 251.2 \text{ k-ft} \\
\text{restoring moment} &= (3 + 2 + 11)(21/2) = 168 \text{ k-ft} \\
\text{Net overturning moment} &= 83.2 \text{ k-ft}
\end{aligned}$$

This requires an anchorage force at the wall ends of

$$T = \frac{83.2}{21} = 3.96 \text{ kips}$$

Since the safety factor of 1.5 for the overturn has already been used in the computation, it is reasonable to consider reducing this anchorage requirement to 3.96/1.5 = 2.64 kips if it is used in the form of a service load. In addition, the wind loading permits an increase of one-third in allowable stress, which may also be used to reduce the requirement. Finally, there are added dead load resistances at both ends of the wall. At the corridor the beam sits on the end of the wall and at the building corner this wall is reasonably firmly attached to the wall around the corner. Thus the real need for an anchorage device is questionable. However, most structural designers would probably prefer the positive reassurance of such a device.

References

1. *National Design Specification for Wood Construction,* National Forest Products Association, Washington, D.C., 1986.
2. *Timber Construction Manual,* 3rd ed., American Institute of Timber Construction, Wiley, New York, 1985.
3. *Uniform Building Code,* 1985 edition, International Conference of Building Officials, Whittier, CA.
4. *Western Wood Use Book,* 3rd ed., Western Woods Products Association, Portland, OR, 1983.
5. *Performance-Rated Panels,* American Plywood Association, Tacoma, WA, 1984.
6. Albert Dietz, *Dwelling House Construction,* 4th ed., M.I.T. Press, Cambridge, MA, 1974.
7. Donald Breyer, *Design of Wood Structures,* 2nd ed., McGraw-Hill, New York, 1986.
8. James Ambrose and Dimitry Vergun, *Design for Lateral Forces,* Wiley, New York, 1987.
9. James Ambrose, *Simplified Design of Building Foundations,* Wiley, New York, 1980.
10. Harry Parker, *Simplified Design of Building Trusses for Architects and Builders,* 3rd ed., Wiley, New York, 1982.

Answers to Selected Problems

The answers given below are for those problems that are marked with an asterisk (*) in the text. As noted in item 4 of the *Suggestions* at the front of the book, confidence in computational work is best gained by a habit of self-checking. The answers given here offer an additional aid in the form of an outside check; however, the reader is advised to attempt to solve the exercise problems first without recourse to the given answers.

In some cases numeric answers are rounded off for accuracy only to the third digit—the usual acceptable degree of accuracy for computations in structural design work. In some forms of structural design problems there are more than a single acceptable answer, and in such cases the answer given is only one of the possible answers.

Chapter 4

4-6-D. 4084 $in.^4$

Chapter 5

5-3-E. $R_1 = 8286$ lb; $R_2 = 17{,}714$ lb

5-4-E. Max $V = 5737.5$ lb at left of R_2; $V = 0$ at $x = 5.625$ ft from R_1

5-5-E. Max M = 14,238 ft-lb
5-5-G. 1.88 ft to right of R_1

Chapter 6

6-3-C. Yes, f_v = 80 psi at 9.5 in. from support

Chapter 8

8-3-A. 1.18 in.

Chapter 9

9-4-A. 8 × 12 or 6 × 14 (6 × 14 has least area)

Chapter 10

10-3-A. 1. 2 × 10 at 16 in.; 2. 2 × 12
10-4-A. 2 × 6
10-5-A. 2 × 10

Chapter 13

13-1-A. 18,480 lb (limited by bolts)

Chapter 18

18-2-A. 21.2 kips

Index